ALFREDO ARCHILLA

The Footprints

OF THE Atoms

A New Paradigm for the Origin of Life

outskirts
press

The Footprints of the Atoms
A New Paradigm for the Origin of Life
All Rights Reserved.
Copyright © 2023 Alfredo Archilla
v2.0

The opinions expressed in this manuscript are solely the opinions of the author and do not represent the opinions or thoughts of the publisher. The author has represented and warranted full ownership and/or legal right to publish all the materials in this book.

This book may not be reproduced, transmitted, or stored in whole or in part by any means, including graphic, electronic, or mechanical without the express written consent of the publisher except in the case of brief quotations embodied in critical articles and reviews.

Outskirts Press, Inc.
http://www.outskirtspress.com

ISBN: 978-1-9772-6293-6

Cover Photo © 2023 wwwistockphotos © 2005 The patriot-News. All rights reserved. [Reprinted/used] with permission.

Outskirts Press and the "OP" logo are trademarks belonging to Outskirts Press, Inc.

PRINTED IN THE UNITED STATES OF AMERICA

Table of Contents

Introduction   i

Prologue   v

Chapter 1: About the Origin of Life   1
    What do we mean by origin? 2
    What do we mean by life?.. 4
    The Paradigm Shift of Life 7
    What do we mean by the Origin of life? 16
       Atomic Origin of Life 20
       Carbon-Based Origin of Life 27

Chapter 2: What is the ID Proposition?   34

Chapter 3: What is the ID Hypothesis?   36
    Discovery Institute.. 38
    Center for Science and Culture FAQs 40
    Inferred ID Hypothesis. 44

Chapter 4: Is the ID hypothesis science?   45

Chapter 5: Is the ID hypothesis scientific?   53
    Does the ID-H meet the SMA criteria?. 55
    The Phlogiston Theory. 58

Chapter 6: Inferred ID Argument   60

Chapter 7: Is the Inferred ID Argument Valid?   65

Chapter 8: Is the ID Hypothesis False?   72

Chapter 9: In Search of the Truth   73

Chapter 10: Testing a Theory False   77

Chapter 11: Falsifying a Hypothesis   80

   Testing Karl Popper's Falsifiability Concept 82
      Blue Winged Unicorns. 88
      Truliviability Concept 94
      Is the ID hypothesis falsifiable? 95

Chapter 12: The Paradigm Shift of Existence   97

Chapter 13: Testing a Historical Event Hypothesis   104

Chapter 14: Comparing Competing Hypotheses   106

   The Comparison Logical Framework.. 108
   ID and Abiogenesis, competing hypotheses?.. 112
      The pyramids of Egypt 116
   ID and Abiogenesis, the only competing hypotheses?.. 118
      The Biological Entity.. 119
      The Biological Entity Argument 124
      Is the BE the ID? 126
      Genetic Manipulation 130
   The Comparison Method 133
      The Explanatory Power of a Hypothesis.. 136

Chapter 15: Is the ID hypothesis a theory?   141

Chapter 16: The Paradigm Shift of Information   143

   What is a thought?.. 143
   What is Information?. 145
      Can you guess?.. 146
      Physical Representation of Thoughts.. 152
      Information tests 154

Chapter 17: An Alternate Argument for the Origin of the DNA Information   160
    The Parietal Art 164

Chapter 18: Genetic Information vs. the Cell   166

Chapter 19: Can information alone create a new cell?   168
    Hello World!.. 171

Chapter 20: Genetic Information vs. Human Beings   173

Chapter 21: Is the cell processing information?   175

Chapter 22: An Argument for the Origin of the Carbon-Based Cell   179

Chapter 23: From Matter to Life   184

Chapter 24: Back to the Origin of Life   188

Epilogue   194

Important Definitions   196

Logic 101   205
    What is logic? 205
        What is a logical argument?.. 207
            Argument Examples.. 208
        What is a logical analysis? 209
        What is a fallacy? 210
        Reasoning Modes. 220
            The Beans from the Bag. 225

Bibliography   227

About the Author   238

Dedication   239

Introduction

I always embrace this kind of analysis with great optimism and silent hope that one of these days, humankind will eventually discover the origins of life as we know it.

And it is excellent to learn about the advances scientists make at the forefront of this effort and the disappointing failures and setbacks they encounter, both equally necessary to advance our knowledge about the reality we all live in.

But it is far more interesting to see how an increasing number of well-educated people from different beliefs and dogmas around the world are detouring from their traditional blind faith into the realms of science in the search for answers to relevant questions about our observed reality. Some are motivated by a genuine interest in finding facts and evidence to separate lies from truth. In contrast, others seek confirmation of their rooted beliefs with a clear intention to lie and deceive an audience of people that, for many reasons, have a particular need to perpetuate their ideas. And if you tell them what they want to hear, whether true or false, they will always accept this new scientific work as undeniable truth.

However, the biggest challenge these educated groups of believers have is not to understand the points of view that science disciplines impose to search for those answers but to overcome the deep cognitive biases engraved in their "mind's DNA" by their respective traditions and beliefs.

These inherent biases make them look at reality from a perspective that distorts most of the reasoning processes they use to reach many conclusions.

And although some of the work published by this new breed of scientists is very detailed and well-documented, most of it fails to reach logical conclusions, perhaps driven by their author's cognitive bias. This major limitation effectively impedes their ability to make significant contributions to science.

Unfortunately, our greatest minds are busy working at the front edge of the knowledge wave, preparing the road for future generations. And they don't seem to dedicate enough time to do good peer reviews and effectively filter these scientific proposals, leaving a space in our society where deception, delusion, and misconceptions live and prosper.

However, I have found it very difficult to ignore the above scenario as it unfolds in front of my own eyes without feeling a moral and ethical responsibility to future generations to seek to differentiate, to the best of our knowledge, between a truth and a lie.

And although I'm trying to survive my day-to-day endeavors, I have committed to reviewing a few exciting topics that have been hostages of old traditional points of view. To help those who want to know the truth but don't have the time and knowledge to analyze these materials effectively.

In this respect, too many topics and publications on different subjects exist. But for many reasons, I have chosen to address the case of the origin of life on Earth. This topic has been stagnant for a couple of decades without a clear direction on where to go next.

While researching this topic, I found ignoring some published work virtually impossible. Especially the intelligent design hypothesis documented in the *Signature in the Cell* book about the DNA Evidence for

Intelligent Design written by Dr. Stephen C. Meyer (Meyer, 2009), Senior Fellow of the Discovery Institute[1] and Director of the Center for Science and Culture of Seattle, WA.

I found his proposal well aligned with the main topic of this work, but also very intriguing, somehow unique in its reasoning approach, and honest in its pursuit of answers to very fundamental scientific questions about the origin of life and our existence here on planet Earth, but above all very well documented, with an excellent history of events associated with its interaction with the scientific community, the general public, the public school system, and our legal system.

And if you have never heard about the Intelligent Design hypothesis, I recommend you gain some basic understanding before reading this material. You will find lots of information on the subject online that will significantly facilitate the comprehension of the information contained herein.

Therefore, the work I describe here is primarily my interpretation and critical evaluation of some of the most predominant topics of science related to the origin of life on Earth.

But this personal interpretation could not be adequately stated and communicated without challenging some of the most basic and relevant aspects surrounding our current paradigms about life, our existence, and other relevant scientific concepts. So, suppose we want to advance our understanding of these critical topics. In that case, we need to be open to the possibility that maybe it is time to consider challenging some of our traditional points of view with a different reference framework.

I also recommend that you glance in advance through the *Important Definitions* and the *Logic 101* sections at the end of the book to become

---

[1] Discovery Institute | Public policy think tank advancing a culture of purpose, creativity, and innovation.

familiar with some important definitions and the topics of logical fallacies and reasoning modes. These will help you gain some basic knowledge to comprehend better the materials presented here.

I hope that all of you can find it interesting, but more importantly, it can help us advance in the coming years one more step to shed some light on this vital topic of the origin of life on planet Earth.

Prologue

Throughout these materials, I will review some of the most relevant aspects of the beginning of life on Earth, but not all of them. I will focus primarily on the topics that I think will have the most significant impact to help us understand the logical implications of our current paradigms, which you will recognize as a recurrent theme throughout the book.

When adequate, I will also take the opportunity to evaluate some topics to challenge current assumptions, definitions, and interpretations that make up our existing paradigm about life on Earth.

You will find lots of constructive confrontations on the topics that I strongly disagree with, as well as very positive remarks on the ones with which I strongly agree. All of this with one end in mind to contribute to advancing our understanding of those important topics.

Chapter 1 is about the "origin of life." When combined, the basic definitions of "origin" and "life" are revised to improve our understanding of their meaning.

Also, I take the opportunity to introduce a new definition of life to reconcile the meanings of life from different disciplines of science under the same umbrella.

Then a comparison between the traditional and the new definition of life is outlined, which opens the need to explore the Intelligent Design hypothesis as a potential explanation for the origin of the carbon-based cells.

Chapter 2 explains what I perceive the Intelligent Design proposition is all about.

Chapter 3 explains some of the challenges I had trying to identify the empirical observation of the Intelligent Design hypothesis and what ID advocates were trying to prove.

Chapter 4 and 5 challenge the generally accepted notion that the ID hypothesis is not science and explains in detail why it is a scientific hypothesis.

Chapter 6 describes what I perceive is the main argument used by ID advocates to try to prove that human beings and nature are the results of the creation of an intelligent designer.

Chapter 7 analyzes the inferred Intelligent Design argument and explains why it is inadequate to prove that the ID hypothesis is true.

Chapter 8 argues why the ID hypothesis is not false, even though its main argument contains logical fallacies and reasoning errors.

Chapter 9 is a short philosophical review of the fundamental nature of science as we seek the truth in our observed reality.

Chapter 10 reviews the importance of testing our false theories to advance science in the pursuit of knowledge.

Chapter 11 analyzes Karl Popper's falsifiability concept and explains why this concept is not practical to determine the scientific nature of hypotheses.

I also introduce the Truliviability concept to help me illustrate and reinforce the conclusion of this analysis.

Chapter 12 introduces a new paradigm shift regarding the non-existence of "something" in our observed reality to help us explain and close the door behind the long-standing question of why it seems so difficult to prove a negative.

Chapter 13 reviews the challenges we encounter when testing hypotheses of historical events, like the one proposed by ID advocates.

Chapter 14 reviews why the Intelligent Design and the abiogenesis hypotheses are not competing for the empirical observation of the genetic information in our cell's DNA. It also explains why these two hypotheses are not the only ones that could explain the presence of this genetic information in the cell's DNA.

It also challenges the practical use of the explanatory power of a hypothesis within the philosophy of science as a comparison tool. And it explains in detail why the confidence level in this method to choose the best explanation is very low.

Chapter 15 will challenge the notion that the ID hypothesis is a theory, as presented and argued, and explain why it is only a hypothesis.

Chapter 16 introduces a new paradigm of the information concept related to our thoughts and how we communicate in our observed reality. It is relevant to understanding the origins of the genetic information in the cell's DNA.

Chapter 17 presents an alternate argument for the intelligent design hypothesis that avoids the logical fallacies and reasoning errors of the original argument presented by its advocates. Also, it reviews the difference between information and the media where information is stored.

Chapters 18 and **19** explain the paradox of genetic information and the cell. It reviews the importance of avoiding the confusion between the correlation and causation for both effects.

It also explains why the genetic information contained in the carbon-based cells can't be the cause responsible for the origin of the carbon-based cell.

Chapter 20 explains why the cause responsible for the emergence of the genetic information in our cell's DNA does not necessarily have to be the exact cause accountable for our presence here on planet Earth.

Chapter 21 reviews the subject of whether or not the cell of the carbon-based living entities has logical processing capabilities to process information.

Chapter 22 presents an argument for the origin of the carbon-based cell based on its information processing capabilities.

Chapter 23 challenges the random environmental condition of abiogenesis hypotheses as a viable one to produce the carbon-based cell. It also presents a new definition for biogenesis based on the principles of the conservation of matter and energy and our knowledge of the biogenesis process.

Chapter 24 summarizes the atomic and carbon-based origins of life. It explains why the intelligent entity cause is a potential explanation of the emergence of the carbon-based cell in our observed reality.

It also makes a general review of evolution's important role in the existence of the wide range of carbon-based living entities on planet Earth and how it relates to the carbon-based origin of life.

CHAPTER 1

About the Origin of Life

So much has been said about the complex and controversial "origin of life" (Shikha, 2022). And I guess we still have many more years to come to write about it, especially when there are so many interpretations of this phrase.

But, when there are too many interpretations or meanings of something, a fundamental reason usually escapes everyone's eyes.

I witnessed this situation many times throughout my professional career while working with teams of talented engineers trying to solve a problem in a manufacturing line. Everyone knew (or thought that they knew) the root cause of the problem, but when asked, all of them had a different explanation about what was causing it.

So what do you do in a situation like this? You scratch your head and flip a coin to choose one of the two most convincing explanations, right?

Well, the answer is no. You need to go back to all the basic assumptions that describe the problem, including what everyone thinks is the problem itself, and use different analytical tools to collect enough data to identify the hidden root cause or causes and determine an adequate solution to the problem. Japanese manufacturing companies have learned to do this very well to maintain the quality of their products and services.

However, based on my experience, more often, identifying the real root cause of the problem relies on a fresh look at the available evidence from a different perspective. This new perspective usually eliminates incorrect interpretations and assumptions that hinder our ability to see the real problem.

The different definitions and points of view that we see today of the phrase "origin of life" comes in the first place from different assumptions and interpretations about what is meant by the word origin, the term life, and the combination of these words in a single phrase.

And although there are many starting points to begin a discussion on a subject like this one, I will go back to the basic definitions of the primary two components of this phrase (origin and life) to analyze them before combining them to determine its meaning.

What do we mean by origin?

When we talk about the origin of "something," we are talking about the identification of an existent cause that was capable of acting upon the things or persons of our observed reality at a given location and time (presence) to derive the effect of that "something" in our observed reality.

Figure 1 below can help us understand the meaning of the origin of "something" as it relates to these essential characteristics.

In the first place, a cause responsible for the origin of something is not a real cause if we cannot directly or indirectly prove its past existence in our observed reality. So causes that lack objective evidence to prove their past existence can't be considered real causes responsible for the origin of "something."

Second, a cause is not an adequate cause to explain a derived effect of something if we can't prove directly or indirectly that it can produce the desired result. So causes that exist but lack the means or mechanisms to generate a derived effect over the things or persons can't be adequate causes responsible for the origin of "something."

Third, a cause is not an adequate cause to explain an observed effect if the cause existed and was capable of producing the observed effect but was not present at the time and location where the observed effect was derived.

Therefore, if we want to prove the origin of "something," we need to provide evidence of the existence, capability, and presence at a given time and location of a cause responsible for deriving from the things or persons in our observed reality the first occurrence of that "something."

Without meeting these basic requirements, it is impossible to prove the origin of "something."

THE ORIGIN OF SOMETHING

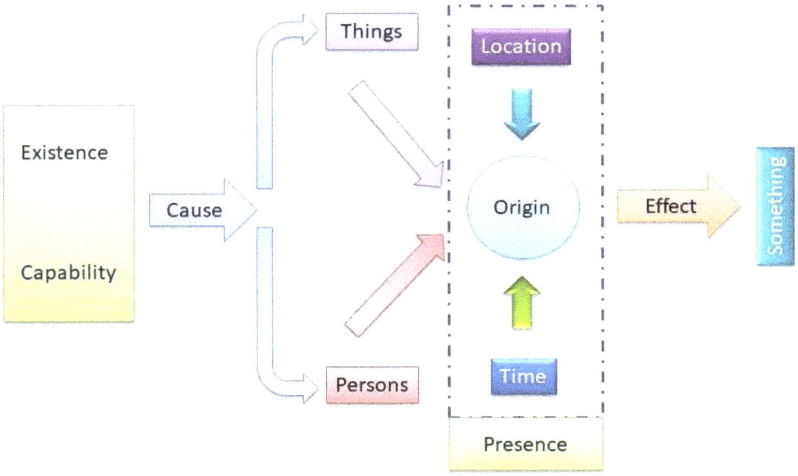

Figure 1 - The Origin of Something

What do we mean by life?

The second part of the equation in the phrase "origin of life" is the word life. Therefore, it is necessary to clearly define and understand what we mean by life to determine its origin.

But "life" is more elusive than "origin" because the word life could mean different things to different people in various science disciplines.

And this imposes an interesting logical challenge to our quest to find the meaning of the "origin of life" phrase. Because it is virtually impossible to find the origin of "something" without having a clear understanding or agreement of what is that "something" that we are trying to find its origin.

And although almost everyone thinks that being alive is the only necessary condition to understand what is meant by the word life, no one can provide a definition that can satisfy the different observed attributes or characteristics of life (Brooker, Widmaier, & Graham, 2008) in all disciplines of science (Science, 2017), and there is probably a good reason for this hidden within the attributes of life that we have chosen to describe this phenomenon.

However, that does not mean we have no clues about this observed phenomenon we call life since we have several definitions of life derived from the metabolic, physiological, biochemical, genetic, thermodynamic, and autopoietic processes (Margulis, 2022); this only means with all probability that the definitions of life that we have chosen to describe this phenomenon are probably biased or incomplete within each discipline of science, which raises a fundamental question about whether or not these are adequate for the observed phenomena that we are trying to describe.

Therefore, under the current scenario, we have to recognize that if the definition of life itself is a challenging task to agree upon, more difficult is to agree on the origin of a life that is not completely defined and understood. The logical implication to science is that we could have as many "origins of life" as definitions of "life" we can come up with.

However, to demystify this inherited challenge, we must choose a starting point.

So, let's choose the most generally accepted definition of life from the biology field (Lim & Dutfield, 2022) as a good starting point. And this is because of a good reason since this is the branch of science that studies the structure, function, growth, origin, evolution, and distribution of living organisms. And it is the one that better describes the kind of biological entity that we are. So, it is easier for everyone to relate to this definition of life rather than other definitions.

Then, staying within the boundaries of what we currently recognize as living organisms, it is generally accepted that to classify an entity as a living organism, the organism;

- Must have cells (basic unit of life), the tissue formed of cells, and organs made of tissues (Cells and organization).
- Must acquire and use energy to maintain their complex living systems (Metabolism).
- Must be able to adapt to their changing environment (Adaptation).
- Must maintain an internal environment conducive to cell metabolism (Homeostasis).
- Must have cells that divide to create more cells to form different tissues and organs (Growth and Development).
- Must have genetic material (DNA) containing the information and instructions to form new organisms (Reproduction).
- Must exhibit change throughout many generations to adapt to its changing environment" (Evolution).

And an organism that does not meet <u>all of the above attributes or characteristics</u> is not considered a living organism.

Note that this imposes a logical **AND** condition of all these seven characteristics to define the state of what we consider an organism that is alive. But at the same time, it opens the door to exploring different definitions of life, which could give us a different set of logical conditions to define the state of what could be considered a living organism.

Now there has to be a reason why we end up with these specific characteristics or behaviors necessary to recognize a living organism. And it seems to me that it has to do with the past and present knowledge that we have gained from the living organisms (carbon-based) that we have studied in nature. And that the more we learn about them, the more characteristics we add to the biological definition of life.

And there is probably a good reason for this since we are biased in that we see our species (Homo sapiens) as the top living biological entity over the surface of planet earth that possesses all of these attributes. So, our definition of life has to be consistent and compatible with our comprehension of the kind of biological entity that we are.

But, as we continue to learn more about our biological complexity, our definition of life will probably continue to move toward this new complexity, equating, in the long run, the future meaning of life to our cumulative knowledge of what defines our species. For example, it seems logical to think, based on the historical evolution of the traditional definition of life, that as soon as we have a better understanding of our capacity to think and how this one relates to our genetic nature, that the next potential candidate to be added to the list of characteristics of life in the biology field will be intelligence.

- Must exhibit the ability to perceive or infer and to retain information as knowledge to apply it toward adaptive behaviors within its environment (Intelligence).

And with the addition of this future characteristic of life, chances are that a few more entities for which we are not sure if they are capable of meeting this new requirement will be left out of the list of living organisms.

If we continue this trend, the biological definition of life will eventually mean the same as human beings. Since the meaning of the word life, in the long run, will be more and more descriptive of our comprehension and cumulative knowledge of human beings, something that could hinder our ability to recognize other life forms in or out of planet Earth.

So, we have to ask ourselves if a different definition of life not derived from the observed attributes or characteristics of carbon-based organisms can help us remove this biased point of view. And at the same time allow us to reconcile all the existing definitions of life in other disciplines of science under the same roof, giving us a new understanding of life and its corresponding origin.

I think that the potential benefits are worth the time and effort of exploring this possibility.

The Paradigm Shift of Life

The new definition of life I propose is based on the assumption that once there was no matter, and now there is, and that everything that exists, including life, is derived from this matter.

It also assumes that life is a fundamental property of all the elements of matter rather than a specific set of observed attributes or behaviors of a particular configuration of some of the elements of matter (i.e., the cell).

And by working on these assumptions, a broader scope of the meaning of the word life is obtained, allowing us to account for all the observed phenomena of living entities in all science disciplines, including but not limited to the already known carbon-based living organisms.

Basic assumptions

The ultimate origin of all observed phenomena in the universe is the origin of all the existing matter that has a mass and occupies space.

All existing matter is made of substances called elements, which have specific physical and chemical properties.

These elements are made of atoms, which is the smallest indivisible particle of an element that conserves the properties of the element.

All atoms are formed of protons, electrons, and neutrons made from other subatomic particles.

Therefore, all things that exist, natural or man-made, living or nonliving, with their corresponding properties, attributes, and behaviors in our observed reality can be explained by the effects of external or internal causes that promote the interaction of the sub-atomic particles of the atoms of the elements of matter.

The new definition of Life

Then, based on these assumptions, let me propose that we use a more fundamental property of the elements of matter to define life rather than a specific set of observed attributes or behaviors of one particular physical arrangement of these elements (i.e., the cell).

Let's define **life as the physical and chemical interactions between nonliving atoms of the elements of matter**.

And let's also name the **basic building block of life: the atom** from which all things that exist in our observed reality, including the cells of all living entities, are formed.

This new definition of life requires interaction between nonliving atoms of different elements of matter, and therefore without this interaction, there is no life.

It also shifts the basic building block of life from the cell, which is a specific physical structure or arrangement of atoms of the elements of matter, to the basic building blocks of the elements of matter, which are the atoms. So essentially, it is a definition of life based on matter's physical/chemical properties (Rocke, 2022).

This change in the basic building block of life brings a different perspective to what is considered a living entity by removing the constraints imposed by the traditional organic definition of life and replacing it with a broader scope of a sustained physical and chemical interaction of the atoms of the elements of matter.

Notice that I used the word sustained because the interaction of the sub-atomic particles of the elements of matter only happens while the physical/chemical reaction is "alive." But once the elements of matter that participate in the reaction are consumed, or the physical/chemical reaction can no longer be sustained for whatever reason, a different stable composition of matter is derived that can't be considered to be alive due to the lack of interactions between the atoms of the elements of matter.

I also use the term "living entities" instead of living organisms to account for all possible life forms that can be explained with this new definition of life.

For example, all living organisms on Earth are formed from oxygen, carbon, hydrogen, nitrogen, phosphorus, and Sulfur atoms (Matter, elements, and atoms | Chemistry of life (article) | Khan Academy).

These elements of matter, under the particular structural arrangement of the carbon-based cells and our environmental conditions, sustain the interaction of the atoms of these elements to exhibit specific attributes or characteristics. These attributes include responsiveness, growth, metabolism, energy transformation, and reproduction. But once these interactions of the atoms cease (for whatever reason), the organism is no longer considered alive.

However, with this new definition of life, it is possible to account for any life form or living entity in any place of the universe under any environmental conditions.

This new life could be formed of other combinations of the elements of matter and exhibit different attributes or characteristics of life, for example, a Selenium-based life form[2].

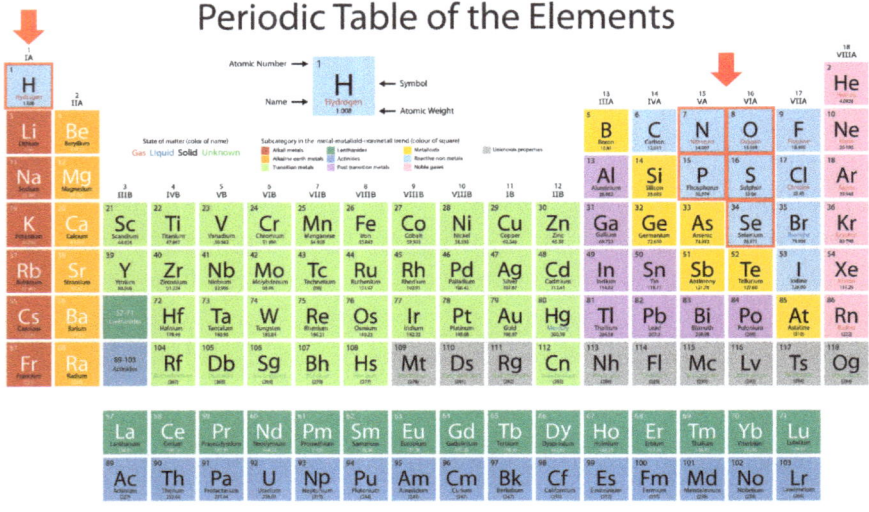

Figure 2 – Selenium Based Life

---

2   Hypothetical example.

This new definition of life also implies that any physical and chemical interaction of the nonliving atoms of any of the elements of matter in our observed reality (not only inside the cell structure of the carbon-based living cells) is alive while the interaction is sustained. This implication brings a whole new set of other life forms or living entities that do not meet the current biological definition of life.

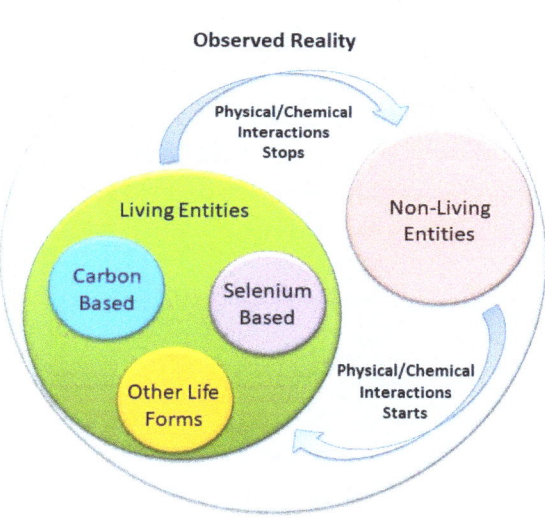

Figure 3 – Living vs. Non-living Entities

Notice, as shown in Figure 3, that there is a continuous flow of entities from the living to the nonliving domain and vice-versa, triggered by the start/stop status of the physical and chemical interactions.

This flow of entities from one domain to the other is based on whether or not the atoms of the elements of matter sustain physical and chemical reactions, for example, when a living entity can't longer sustain its physical and chemical reactions and becomes a nonliving entity, and when nonliving entities are induced to start physical and chemical reactions to become living entities. This continuous flow of entities from one domain to the other can be illustrated by a carbon-based living entity that can no longer sustain its physical and chemical reactions in its cells (for whatever reason) and dies. Which eventually decomposes and is transformed into the nonliving elements of matter from which it was derived. And when a new carbon-based organism is formed from the nonliving elements of matter through a biogenesis process.

So, what is a living entity under this new definition of life?

Living Entities

A living entity is anything in our observed reality that sustains a physical/chemical reaction of two or more atoms of the nonliving elements of matter.

Let's see a few examples of these new living entities;

Any living entity that exhibits one or more (**not all!**) of the seven attributes of life currently recognized in biology (Cells and organization, Metabolism, Environmental Adaptation, Homeostasis, Growth, Development, Reproduction, and Evolution) is considered a living entity.

Also, a living entity under this new definition of life does not need to meet all of the attributes or characteristics of carbon-based living entities (**AND conditions**). It must only meet at least one of these seven attributes (**OR condition**).

And this is the case because each of these seven characteristics of living organisms involves sustained physical/chemical reactions of two or more atoms of the different elements of matter inside the cells to perform their functions.

So, for example, a mule that cannot reproduce meets six of the seven recognized attributes of living organisms; therefore, it is alive.

Ribonucleic acid (RNA) molecules that are not alive may replicate and mutate in test tubes under the right environmental conditions by sustained physical/chemical reactions of the atoms of different elements of matter. Therefore, they are alive under the right environmental conditions.

A candle flame can "metabolize" its organic waxes formed from the elements of matter and the surrounding oxygen by a sustained chemical reaction to produce carbon dioxide and water. It is also capable of growing; therefore, it is alive.

A glass of water reacting with a teaspoon of sodium bicarbonate is a living entity under this new definition of life since it is a sustained physical/chemical reaction of the atoms of different elements of matter. But once the chemical reaction is over, it can no longer be considered alive.

Crystals grow and develop in different shapes and colors through nucleation that results from the sustained chemical reactions of ions, atoms, or molecules of many common chemicals formed from various elements of matter (Helmenstain, 2017). Therefore they are considered to be alive and growing while the physical/chemical interactions of the atoms are sustained.

Any substance or material that experiences oxidation or reduction also triggers sustained chemical reactions from the interaction of the atoms of different elements of matter (Helmenstine, Oxidation Definition and Example in Chemistry, 2020), therefore, is considered alive.

A virus (Segre, 2022), considered to be a nonliving matter when isolated, becomes alive when it infects a host cell by activating its reproduction mechanism and initiating sustained physical/chemical reactions of the atoms of different elements of matter to reproduce; therefore it is considered to be alive while it is reproducing inside the cell.

An automobile can also sustain chemical reactions of the atoms of the elements of matter inside its combustion chambers to derive the necessary energy to achieve its functions. Therefore while the engine is on, the automobile is alive.

A cell phone (Mischa, 2016) is also alive while its battery power is consumed because its energy is derived from the chemical reactions of the elements of matter (Ni, Cd, Li, etc.) inside its electrochemical cells.

In summary, as per this new definition of life, any natural or man-made system that sustains physical/chemical interactions of the atoms of the elements of the matter is considered to be alive.

This new definition also sheds light on the fundamental patterns of the forms and the functions of all carbon-based living entities, which are essentially the same because they are derived from the same basic structures and physical characteristics of the atoms of the elements of matter from which they are formed.

So, what is a nonliving entity under this new definition of life?

Nonliving Entities

Nonliving entities are all things that exist in our observed reality that do not sustain a physical/chemical reaction of two or more atoms of the nonliving elements of matter.

For example;

All the elements of matter (Pauling, 2022) are not alive since, as stand-alone elements, they have no physical or chemical interactions with other elements of matter.

All living organisms that can no longer sustain a physical/chemical reaction of the atoms of the elements of matter in their cells (die) are nonliving entities.

An (RNA) molecule, under environmental conditions unsuitable for replication, is a nonliving entity.

A wax candle that is not lit is a nonliving entity.

Crystals that can't longer grow and develop due to the lack of the necessary minerals and substances that support their nucleation process are nonliving entities.

Stable substances or materials no longer experiencing oxidation or reduction, like a glass of water or a stainless steel metal part, are nonliving entities.

A virus (Segre, 2022) in its inactive state, waiting to kidnap a host cell to initiate its reproduction, is a non-living entity.

An automobile whose engine is not running is a nonliving entity.

An off-cell phone is nonliving because no energy is drawn from its battery. Therefore no sustained interactions between the atoms of the elements of matter are happening.

In summary, per this new definition of life, any natural or man-made system that does not sustain a physical/chemical reaction of the atoms of the matter's different elements is considered a nonliving entity.

What do we mean by the Origin of life?

When we read the "origin of life" phrase in many publications and scientific work, almost everyone interprets this as the origin of carbon-based living entities, which exhibit the following attributes or characteristics, essentially equating the origin of the life phrase to the origin of the carbon-based living entities.

- Cells and organization
- Metabolism
- Environmental Adaptation
- Homeostasis
- Growth and Development
- Reproduction
- Evolution

And this is almost always the case because most of us are biased with our impressive characteristics as carbon-based living entities, and it becomes easy for us to ignore the possibility of other life forms not based on carbon.

Now, the fact that we have identified only carbon-based living entities in our observed reality is primarily influenced by the definition we have given to life, which aligns with the attributes or characteristics mentioned above.

So, in essence, we have been looking for life through a "1-inch PVC pipe", ignoring the possibility that a different definition of life could give us a broader scope to look for other living entities not only here but elsewhere in the universe.

After all, it seems almost illogical to think that with all the different elements of matter in the universe (maybe some unknown ones), only one possible combination and structure of matter (the cell) can exhibit life attributes or characteristics. But, the fact that we have only found one (carbon-based life) doesn't mean or conclude that there is only one possible life form. It probably means that due to our short sight, we have only been able to recognize only one.

Now, per our previous conversation about the meaning of the word origin and the meaning of the word life, when combined, it means that the origin of life is a place at a given moment in time where an existent and capable cause acted on the things of our observed reality to derive the effect of life.

But the derived effects of life now have two different potential causes for its existence. One is from the newly proposed physical and chemical interactions between nonliving atoms of the elements of matter, and the other is from the traditionally recognized seven attributes of life that carbon-based cells exhibit.

And, because the two definitions of life are different, as I commented before, we should expect the origin of each one of them to be different too.

Well, that sounds great, but how do we know which one of the two possible origins of life is the "true" origin of life?

That is an excellent question. But, like everything else in science, it is relative to the assumptions used to explain the observed phenomena. However, let me ask if it matters or makes any difference. I say this because it depends on how helpful is any given definition or concept when it comes to our need to apply a piece of knowledge in the solution of the challenges that we face for the survival of our species.

In other words, does it matter which one of the two definitions of life (atomic or carbon-based) is the "true one" as long as each one has practical uses in our quest for the survival of our species?

The above situation is similar to using and applying traditional Newtonian and emerging quantum physics. Both are useful in describing the motion of objects in the space-time dimension. But depending on the problem we are trying to solve, one definition is more practical.

However, I will be surprised to find out that both definitions of life have the same level of supporting evidence when determining its origins. In other words, which of the two definitions of life has more objective evidence to support an existent, capable, and present cause that acted upon the things of our observed reality at the time and location where the first effects of life were derived?

To answer this question, we will have to take a closer look at the "origin of life" for each definition to find out which is the one that has the most supporting evidence for its observed attributes or characteristics.

As you can imagine, this is not an easy task to complete because we are dealing with historical events for which there is little or no objective evidence of the causes that triggered them.

However, based on our cumulative knowledge of the different disciplines of physics (Zimmerman A. J., 2019), we can use our abductive reasoning mode to infer possible causes that can derive these effects of life from the things of our observed reality. And the inferred cause with the most supporting evidence is the best possible explanation we have in the subject with our current state of knowledge. And I say the "best possible explanation" because when we use our abductive reasoning to infer possible causes of past events, there is no 100% guarantee that the identified cause was responsible for these past events. But at least it gives us an explanation to the best of our knowledge for now.

So, we should seek to answer the following questions to determine the most probable cause of the origin of life;

ABOUT THE ORIGIN OF LIFE

What kind of evidence do we have of a cause capable of acting upon the things of our observed reality at the location and time when the first physical and chemical interaction between the non-living atoms of the elements of matter happened?

And,

What kind of evidence do we have of a cause capable of acting upon the things of our observed reality at the location and time when the first cell of the carbon-based living entities was derived from elements of matter?

As a guideline to track the progress in answering both questions, I propose using the following chart. This chart will help us identify the potential cause for each definition of life and replace the question marks with "Y" or "N" according to the availability of evidence that can support the observed attributes of both definitions of life.

The Origin of Life

Definitions of Life	Atomic	Carbon Based
Cause	?	?
Existence	?	?
Capability	?	?
Location	?	?
Time	?	?

Evidence = Y   No Evidence = N   Undefined = ?

Figure 4 - The Origin of Life, Atomic and Carbon-Based

Atomic Origin of Life

Cause

There is no shortage of potential explanations (theories and hypotheses) from our cosmologists and physicists about the universe's origins (Kumar, 2022). All of them are based on mathematical models with different assumptions and constraints (singularities, black holes, dark matter, energy, etc.) that draw the lines on the boundaries of our knowledge about the things we can't explain. But all of them, one way or the other, are trying to answer the same fundamental question about how all the matter that has mass and occupies space came into existence in our universe.

However, the most generally accepted theory today, with known limitations or constraints, that has the most compelling evidence for the existence of all this matter came from the observation that all other galaxies are moving away from ours at great speed in all directions as if the action of an expanding force or a Big Bang (Bortz, 2014) is driving them away from us.

This explanation of the origin of our universe is supported by the use of physical laws and by the observed effects that this ancient expanding force still has on the things around us (Greshko, 2017) to explain the origin, development, and nature of our universe. And based on this empirical evidence (Panda, 2020), we have inferred the existence of this ancient cause that we name the Big Bang as the one responsible for the origin of all the existing matter in our universe.

The proposed atomic definition of life aligns with this Big Bang theory by recognizing a time when no matter, space, energy, radiation, or light existed as we define these in the space-time dimension of our observed reality with our laws of physics. And now all of these things exist. Therefore based on the knowledge that we have about causation, it makes sense to think that "somewhere," a capable cause derived the effects of all these things from "something" (a singularity?) that was in existence at the origin of our universe. And these effects were responsible for the presence of all matter and the emergence of living entities on planet Earth.

Existence, Capability

However, this happened so long time ago, and there were no witnesses from our species around when the event or events that unfolded during these early days of the existence of our universe happened. Therefore we still can't comprehend very well and can't even describe very well what that "thing" was or "where" it was when our universe originated. But based on the evidence of the effects left behind by this cause, we can infer that this cause existed and could derive all the elements of matter that have mass and occupy space in our universe.

Redshift of light

During the expansive force of the big bang, different galaxies, objects, and other bodies were expelled at various speeds causing them to travel further distances.

This difference in distances from galaxies as they continue to drift from each other causes the wavelength of the light that they emit or Redshift (Rhee, 2013) as perceived on earth to experience the Doppler Effect or increase in wavelength (Byjus, 2022).

Therefore, the longer the wavelength of light received from a given galaxy, the farther it is from the earth; this is the basis of Hubble's Law that correlates the recessional velocity of a galaxy as it moves away from earth.

This evidence supports that galaxies went through a significant expansion at the beginning of our universe and that they are moving away from each other as measured by the redshift of light.

Cosmic background radiation

The premise of cosmic background (electromagnetic) radiation is that the interaction of matter and light in the universe created cosmic radiation in the form of galaxies and stars when the big bang event happened, giving us an early view of the universe right after the big bang. The density of this original radiation has been decreasing since its inception due to the expansion of space after the occurrence of the Big Bang.

This decrease in density has been confirmed through measurements of the redshift in the electromagnetic radiation of the universe by the NASA Cosmic Background Explorer (COBE) launched in 1989, and the Wilkinson Microwave Anisotropy Probe (WMAP) launched in 2001, which detected the existence and pattern of cosmic background radiation across the globe.

Hence, the existence and the variation of the cosmic background radiation measurements of the universe support the presence of an earlier universe stage right after the Big Bang event (Panda, 2020).

The abundance of primordial elements

The primordial elements of matter are the most abundant elements of matter in the universe. These are hydrogen, helium, deuterium, and lithium.

These primordial elements are the basic "building blocks" of all other elements of matter, which are formed through a process called Nucleosynthesis[3]. This process creates more complex atoms of different elements from one or two simple atoms of these primordial elements by adding or removing protons or neutrons through large-scale nuclear reactions like the ones in progress in the Sun and other stars in our universe.

Since discovering these primordial elements in the sun, further investigations have revealed their existence all over the universe; this supports the theory of the Big Bang as the responsible cause of generating these primordial elements of matter, from which all remaining matter was formed and expanded to fill the universe (Panda, 2020).

Structure and distribution of galaxies

The structure and distribution of galaxies, galaxy clusters, and large bodies in space also support the existence of the Big Bang because it follows some organization in its density pattern.

This distribution confirms that an expansive force accelerated galaxies and other bodies in the universe following a hierarchical pattern in which smaller structures formed first and later merged into large structures.

It also shows that planets and other bodies have oval shapes due to the centrifugal forces that caused bulging at the equatorial planes and centripetal forces that caused flattening at the polar planes (Anderson, 2015).

Therefore, the Big Bang theory provides a plausible explanation of the structure and distribution of bodies in space.

---

[3] https://www.merriam-webster.com/dictionary/nucleosynthesis

Existence of pristine clouds

Pristine clouds are formed from gasses of atoms that are lighter than primordial atoms, such as lithium, deuterium, and hydrogen. Astronomers established their existence before their discovery, which has been confirmed to exist through analyzing the spectra of distant quasars and galaxies (Canetti, Dewes , & Shaposhnikov, 2012).

The Big Bang nucleosynthesis accurately predicted the existence of such clouds in space before the universe's expansion and emergence of bodies; therefore, it provides compelling evidence that the Big Bang took place in the past horizon.

Place

Although the Big Bang brings the image of a giant explosion like the one of a detonating bomb that emanates lots of debris all over the place from "ground Zero," the Big Bang is not an explosion.

The Big Bang is the very first time that the universe that we see in our observed reality could be described in terms of the particles, antiparticles, and radiation that started to expand and cool from this initial state or conditions according to the laws of General Relativity (Zimmerman A. J., 2020). This initial state eventually formed the stars, galaxies, and the large-scale structures we see today.

But as far as we know, there was no particular place where the origin of this universe started. There is no "burned black hole" in the fabric of the universe that you can point at and say, "Aha, this is the place where it happened!"

On the contrary, all evidence suggests that the Big Bang occurred everywhere at once.

And this is the case because the universe appears to have the same properties everywhere and looks the same in all directions; this means that the Universe is homogeneous and Isotropic. And you don't get a Universe with those properties from an explosion.

In an explosion, faster-moving particles travel the farthest away, showing a more dispersed pattern. In other words, greater distances would appear to have fewer galaxies per unit of volume. But this is not the case in our universe.

Therefore, the fact that the Universe is homogeneous and isotropic tells us that the Big Bang happened simultaneously at all locations equally (Siegel, 2016).

Time

Alexander Friedman initially determined the universe's age to be 10 billion years in 1922 using Albert Einstein's relativity. But it was later revised to be 13.82 billion years, give or take 21 million years, by compensating for the effect of the actual expansion rate of the universe (Hubble constant) by measuring the cosmic microwave background of the universe with the use of the Planck space telescope in 2013 (IOP, 2022).

Conclusion

Therefore, based on the above evidence and our knowledge about causation, we can conclude that there was a capable cause responsible for deriving the effect of the expansion of our universe in all directions approximately 13.8 billion years ago. And that this effect (Big Bang) was responsible for all things in our universe, including the basic building blocks of life (the atoms), the elements of matter, and their eventual physical and chemical interactions that sustain all living entities of our observed reality.

Below is the updated Figure 5 reflecting the progress made regarding the origin of the atomic life.

The Origin of Life

Definitions of Life	Atomic	Carbon Based
Cause	Big Bang	?
Existence	Y	?
Capability	Y	?
Location	Everywhere	?
Time	13.8 BY	?

Evidence = Y　　　　No Evidence = N　　　　Undefined = ?

Figure 5 - The Origin of Life, Atomic Update

Carbon-Based Origin of Life

When we talk about the carbon-based origin of life, it is essential to realize that there is a dependent causal relationship between the origins of all living entities and the atomic origin of life. Because all existent living entities (including carbon-based ones) are formed from the atoms of the elements of matter, without them, it is impossible to explain their existence.

In other words, we can't explain the existence of the carbon-based cell without first explaining the existence, structure, and configuration of the oxygen, carbon, hydrogen, nitrogen, phosphorus, and sulfur atoms inside the cell since these are arranged in a specific structure and configuration that support the functions of the cells.

In that sense, although we still don't know all the details about the structure and functionality of the cell at the atomic level, I will venture to say that eventually, it will be possible to explain it in terms of the interaction of the physical and chemical reactions of the atoms of these elements of matter.

However, in the same way that it is not possible to determine the atomic origin of life without explaining the origin of the atom (Big Bang), it is not possible either to determine the carbon-based origin of life without explaining the origin of the cell. But the good news is that if we can find the origin of the carbon-based cell, we should also have found the origin of all carbon-based living entities formed from these cells.

Cause

But how did those elements of matter get organized on the specific form and structure that enabled them to form a carbon-based living cell? Is there a capable cause responsible for acting on the existent matter to derive the carbon-based cell from which these seven attributes of life are derived?

THE ORIGIN OF THE CARBON-BASED CELL
Single Cause

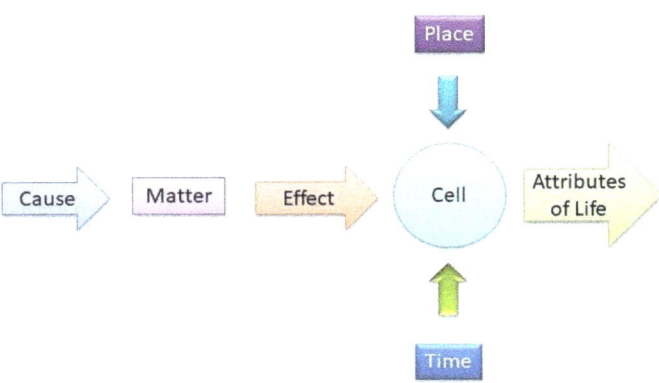

Figure 6 - The Origin of the Carbon-Based Cell, Single Cause

Well, the fact that the cells of all carbon-based living entities are made from those elements of matter tells us that there was a cause responsible for rearranging those elements to form their cells. However, it is also quite possible that the origin of the cells of all carbon-based living entities resulted from a wide range of dependent causal events in different lines of action at various locations and moments in time.

THE ORIGIN OF THE CARBON-BASED CELL
Multiple Causes

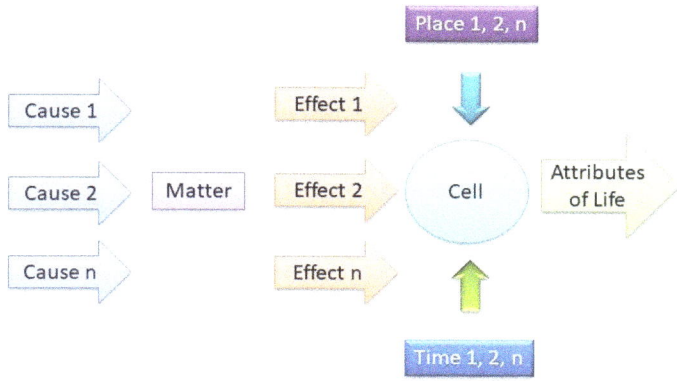

Figure 7 - The Origin of the Carbon-Based Cell, Multiple Causes

But we still don't have a clear-cut answer to this question. We don't know if it was a single cause or multiple causes responsible for the specific structural arrangement of the elements of matter in the cells of the carbon-based living entities.

However, we inferred potential causes to explain some (but not all) of the dependent causal events necessary for the progression of carbon-based living entities since the emergence of our universe up to this day.

Amongst which the most commonly known or relevant ones are;

- The Big Bang: The origin of our universe
- Abiogenesis: The random emergence of life from inert matter
- Panspermia: Simple life forms from outer space
- Natural Evolution: The origin of species
- ET Visitors: Complex life forms visiting Earth
- Biogenesis: Only life can give life

- Technical Evolution: The future origin of species
- Intelligent Design: Creation of Human beings and nature

Figure 8 shows one of many potential representations of a generic timeline of their occurrences.

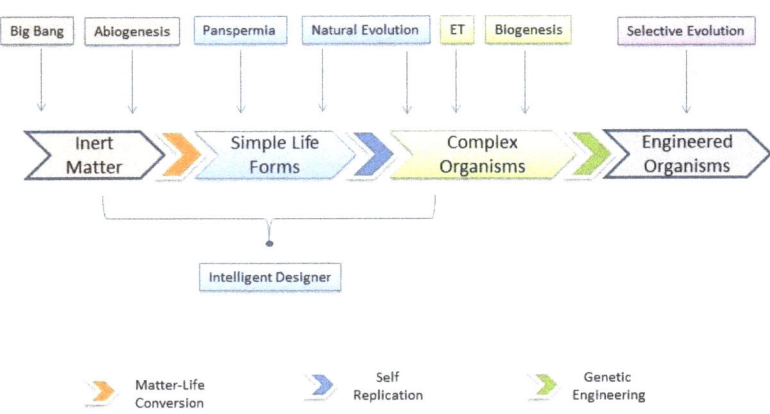

Figure 8 – Carbon-Based Life Roadmap, Single and Multiple Causes

But, if we want to find the origin of the carbon-based cell, the challenge we have in front of us is to determine which cause or causes are the most likely ones responsible for deriving this cell from the elements of matter.

A quick review of these causes shows that not all are pursuing an explanation for the emergence of the cell of carbon-based living entities.

- The Big Bang Theory is an explanation for the origin of our universe and is not pursuing an explanation for the origin of the cell of all carbon-based living entities.

- Panspermia is the belief that life on earth was derived from "seeds" of extraterrestrial origin. But this hypothesis does not explain the origin of the cell of carbon-based living entities.
- The Theory of Evolution assumes that the cells of carbon-based living entities were already in existence and explains how populations of these entities change their characteristics over time to adapt to a changing environment. It also explains how life diversifies to create new species, but it does not explain or provide any evidence of the origins of the cell of carbon-based living entities.
- Although there are a lot of subjective evidence and many "conspiracy theories" about extraterrestrial visitors from other places of the universe as the responsible cause for the existence of life on our planet, none of them, as of this day, provides good objective evidence to prove that these were responsible for the origin of the carbon-based cells.
- Biogenesis is the process by which life is derived from pre-existing life through reproduction. But it does not explain the origins of the first carbon-based living cell; it assumes the existence of these cells and describes their ability to reproduce.
- Technical Evolution is evolution on the hands and guidance of our cumulative knowledge about carbon-based living entities through the use and application of technologies like Artificial Intelligence, Quantum Computing, and genetic engineering. It will eventually explain how our species will evolve in the future through the use of knowledge as the driving force, but it does not go back in time to explain the origin of the cell of carbon-based living entities.

Only abiogenesis hypotheses are pursuing potential explanations for the emergence of the carbon-based cell of living entities in our observed reality.

- Abiogenesis hypotheses attempt to demonstrate a process that could explain how simple organic compounds and cells emerged randomly from physical matter under primitive environmental conditions. So, this one is probably a good candidate that could eventually explain the origin of the carbon-based cell. However, there is still no consensus amongst the various available hypotheses on achieving this. Therefore there is no confirmed process capable of deriving the effect of the carbon-based cell from the elements of matter and no supporting evidence about a potential responsible cause or causes.

On the other hand, the Intelligent Design hypothesis is pursuing a potential explanation for the emergence of humans and nature on planet earth. As such, it is inevitable to avoid explaining the emergence of the carbon-based cell because human beings and nature are formed from these cells.

- Intelligent Design is a relatively new hypothesis that attempts to explain the origins of human beings and nature in the hands of an intelligent designer through the association of our intelligence attribute and the presence of the genetic information contained in our cell's DNA.

This hypothesis raises an interesting question about whether or not the presence of genetic information in our cell's DNA is adequate evidence to prove that the cause responsible for the origin of the carbon-based cell and the eventual emergence of all carbon-based living entities was an intelligent designer.

To determine if this was the case, we must evaluate their hypothesis, understand their argument, and decide whether it is valid, with supporting evidence for the existence and capability of this intelligent cause in a particular location and moment in time.

Therefore, the following chapters of the book are dedicated to discovering if this intelligent designer could have been the responsible cause for creating the carbon-based cell and the eventual emergence of these living entities.

CHAPTER 2

What is the ID Proposition?

After reviewing most of the available online materials about Intelligent Design, I have concluded that its advocates are proposing that the cause responsible for the origin of human beings and nature on Earth was an intelligent entity.

They propose this cause by exploring the connection between our ability as an intelligent entity to create large amounts of information and the intentional creation by an Intelligent Designer (Cherry, 2022) of the large amounts of genetic information contained in the cells' DNA of carbon-based living entities.

Unfortunately, as with all other scientific propositions in which very little or no evidence of the original cause or causes remains as of this day, this one does not rely on experimentation and tangible test results, as in other disciplines of science that we are more familiar with. And because of this, it is challenging to prove or disprove it within the traditional evidence-based scientific framework.

Instead, the intelligent design proposition relies primarily on prior and new research on the subject matter, logical assumptions, and available evidence from other potential causes that could explain the empirical observation of the genetic information in the cell's DNA.

However, like any other scientific proposition, it is not exempt from a detailed logical reasoning analysis to test its arguments, premises, and conclusions to determine if it accurately describes our observed reality.

CHAPTER 3

What is the ID Hypothesis?

After reading some of the Intelligent Design materials, I still wonder if their advocates have spelled the officially proposed intelligent design hypothesis somewhere word by word. And if so, where is it documented?

I say so because many bits and pieces of information are everywhere, worded in so many different ways that they need to be carefully assembled to understand the Intelligent Design hypothesis clearly.

At first glance, it is difficult to determine if the ID hypothesis is trying to explain the emergence of human beings and nature, the carbon-based cell, the genetic information in their DNA, or all of the above. And considering that all of these events could have different but related causes or explanations in the same causal line of action necessary to explain the existence of the carbon-based living entities, it is essential to make a clear distinction.

This lack of adequate definition of what is being proposed adds confusion. It makes you wonder briefly if this was a strategic decision with a specific purpose in mind or an overlooked/involuntary action from its proponents.

But regardless of the reasons that they had for it, it leaves the reader with no option other than to make some assumptions and educated guesses about what is being proposed.

ID advocates may not have defined their proposed hypothesis very well, but just like everyone else trying to convince others or prove something, they have used words to build their arguments to make their points. And these arguments, like any other argument we make in our day-to-day life, can be tested through logical reasoning, which has always been an excellent defense against lies, deception, and errors that attack the foundation of truth.

Using logical reasoning to examine the available Intelligent Design materials will help us determine if Intelligent Design advocates have a sound scientific hypothesis with a valid argument.

But without an easy-to-find hypothesis about Intelligent Design, I have made some inferences to extract "what I perceive" as their proposed hypothesis.

While doing so, I have tried to avoid any personal cognitive biases I may have as much as possible. And I have based the analysis and the proposed explanations only on the available information.

Of course, there is always a possibility that I can be biased or even wrong in my interpretations. But this may be due to the lack of knowledge and information on a given subject or inadequate reasoning, but not because I'm willfully trying to be dishonest or unclear about any particular subject.

But if somehow I have made a wrong interpretation and reached a false conclusion, I can only hope that someone will take the time to clarify the subject in its due time.

Discovery Institute

From the Discovery Institute's Center for Science and Culture (About, 2022), I learned that part of its mission is "***...to advance the understanding that human beings and nature are the results of intelligent design rather than a blind and undirected process.***"

And if I understood well this part of their mission, they are trying to teach others or make others realize or understand "*something*." But the "*something*" they are trying to make others understand does not seem clear enough or well defined in its mission. If anything, it seems more of a subliminal message that needs to be decoded. Let's see why.

If you read what they are trying to advance the understanding, there is an implied assumption that "**human beings and nature are the results of...**" certain actions. But the action that provided the referenced results (human beings and nature) is not adequately identified; this is the case because stating that human beings and nature are the results of "**intelligent design**" does not specify the action that was carried out for human beings and nature to become the result of such action. And this is the case because, in our language, intelligent design is not a verb, and it does not carry any action associated with it. Therefore, you must fill in the gaps between words to clarify the communicated message.

In this case, I can infer from the context in which the words "**...the result of**..." are used that the implied action being alluded to is the act of creation.

Now an "**intelligent design**" by itself can't carry an action to create something since no matter how good a design is, it is only a blueprint, a specification, or a depository of information. Therefore, carrying out the act of creation that is being alluded to requires the intervention of some capable entity with the necessary knowledge, skills, and resources to carry out such an act.

In this case, we can assume, without too much risk of being wrong, that the design or blueprint referenced in this mission is the intellectual product and effort of some intelligent entity capable of producing such a design.

And with no other information available, what is being proposed likely is that the intelligent designer who created this blueprint or design had a purpose for doing so and that it was also the responsible entity who created human beings and nature.

The second part of the mission compares the action of creation from the proposed intelligent designer through the use of the adverb "**rather**" with what is referenced as "…*a blind and undirected process*…", but there is no explicit reference in the mission to what is this blind or undirected process."

However, after reviewing most of the available materials, it seems evident that the implicit "*blind or undirected process*" that is being referenced could be the abiogenesis processes about the beginning of life on Earth or the evolution process of the species since both of them are random undirected processes, this gives us an early hint about the main argument that ID advocates will use to prove their hypothesis by comparing it to other known abiogenesis hypotheses about the beginnings of life on Earth and the theory of evolution.

In summary, we can infer from the mission of the Discovery Institute's Center for Science and Culture the following information:

First, there was or maybe still is a capable, intelligent designer.

Second, this intelligent designer aimed to create a design for human beings and nature, so he did.

Third, based on this design, human beings, and nature then became the direct result of the action of creation by this intelligent designer.

Fourth, human beings and nature couldn't have resulted from a natural random process like abiogenesis or evolution.

Therefore although it is not stated in these specific words, in my opinion, a more complete and easy-to-understand mission for the Discovery Institute CSC could be:

"...to advance the understanding that human beings and nature are the results of the act of creation of an intelligent designer, and not the result of natural random processes like abiogenesis or evolution."

Center for Science and Culture FAQs

In the FAQ (Frequently Asked Questions) section of the CSC website, there is a direct question related to this particular subject:

1. What is the theory of intelligent design?

    "The theory of intelligent design holds that certain features of the universe and of living things are best explained by an intelligent cause, not an undirected process such as natural selection."[4]

It is interesting to note here that there is a difference between the scope of the Discovery Institute's mission and the scope of the answer to this question, which both address the same fundamental need for information about what is the Intelligent Design hypothesis.

The Discovery Institute's mission's main scope is to explain the origins of human beings and nature. In contrast, the scope of the answer to the above question aims to explain *"certain features"* of the universe and living things.

---

4   https://www.discovery.org/id/faqs/

More specifically, the mission of the Discovery Institute states that human beings and nature are the creation of an intelligent designer and not the result of an abiogenesis or evolution process. While the theory of intelligent design, as explained in the above answer, states that certain features of the universe (not all of them) and living things are the creation of an intelligent designer and not the product of the natural selection evolution process.

So, is the ID hypothesis trying to explain the origins of human beings and nature, or only certain features of the universe and living things, or all of the above?

By logically comparing the whole population of human beings and nature and all the features of the universe and living things, there seems to be an implied exclusion of some features that were not part of the creation process of the intelligent designer but the result of other unidentified causes. And this seems to be a gray line that was not adequately addressed or identified in the available ID materials, which makes it difficult to differentiate between both referenced populations of features.

But to simplify matters and move forward, I will ignore this logical exclusion and work on the assumption that the ID hypothesis is aimed to prove that most, if not all, of the universe, its life, and nature as we know it is the creation of an intelligent designer and not the product of natural causes like abiogenesis or the evolution process.

On the other hand, in the answer to this question, there is also an implied comparison between the proposed Intelligent Designer cause and the Natural Selection cause in the words "...**best explained by an intelligent cause, not an undirected cause as natural selection**.", this makes you wonder at this early stage of the game, which is those "...*features of the universe and living things*" that will be used as a comparison gauge between these two causes.

In earlier versions of published online materials on the CSC website (probably not available today), I was able to find the following interesting statements:

1. ID is *"...an evidence-based scientific theory about life's origins-one that challenges strictly materialistic views of evolution."*

Through all the available ID materials, most of the arguments presented by its advocates aim to challenge the materialistic points of view of abiogenesis hypotheses and some aspects of the theory of evolution. Still, we must remember that these two subjects have a big difference.

We can easily argue any hypothesis about the beginnings of life on Earth since those are still hypotheses with limited evidence. But it is much more difficult to argue against the vast body of evidence collected about the species' natural evolution process.

So, the above statement makes you wonder if the author meant materialistic views of abiogenesis rather than evolution since abiogenesis and evolution are two different scientific concepts (Top Differences Between Darwinism And Neo-Darwinism, 2019);

Keep in mind that no abiogenesis hypothesis is trying to explain the evolution process of the species, and the species' evolution theory does not attempt to explain the origin of life on planet Earth.

2. *"Either life arose as the result of purely undirected material processes or a guiding intelligence played a role. Design theorists favor the latter option and argue that living organisms look designed because they really were designed."*

Here, the author presents the two extensively debated options throughout available ID materials for life to arise; first, the undirected abiogenesis processes on **HOW** life arose from inert matter. And second, the intelligent designer **WHO** was responsible for creating human beings and nature.

Notice, however, that the second part of this argument commits a reasoning error or **Circular Reference Fallacy** because its premise *"living organisms look designed"* provides no independent ground or evidence for the conclusion *"because they really were designed,"* making the favorite option of design theorists an unsupported logical conclusion through the use of this argument.

3. *"Based on our knowledge of what it takes to build functionally-integrated complex systems — that intelligent design best explains the origin of molecular machines within cells. Molecular machines appear designed because they were designed."*

Here ID advocates propose that an intelligent designer created the origins of molecular machines within the cells. And once more, the author commits a reasoning error or Circular Reference Fallacy when he says, *"…molecular machines appear designed because they were designed."* The premise that *"molecular machines appear designed"* provides no independent ground or evidence for the conclusion *"…because they were designed"*, making the conclusion of this argument an unsupported logical conclusion through the use of this argument.

Inferred ID Hypothesis

As I said, no clearly stated or identified hypothesis exists throughout the ID materials. And the scope of this hypothesis also needs a better definition since it seems too wide to grasp.

But up to this point, I can infer from the different arguments aimed to prove the same point about the creation of human beings, nature, the universe, and living things that the proposed ID hypothesis or the conclusion that their advocates are trying to prove can be summarized in the following statement:

"Human beings and nature are the results of the act of creation of an intelligent designer, and not the product of random natural causes like abiogenesis and evolution."

As we move forward, the challenge is finding the necessary information to infer the logical argument used by ID advocates to try to prove this hypothesis true.

CHAPTER 4

Is the ID hypothesis science?

One of the most heated debates I have seen regarding the ID hypothesis is whether or not this one is science (Our Definition of Science, 2022). This debate made it from the desks of its proponents to the desks of its opponents, to public forums, to social media and newspapers, to the classrooms of schools, and to the courts of some of the states of our nation. And yet there still seems to be an additional need to discuss it, at least from my end.

The most relevant event in this regard was in 2005 in the federal court in Harrisburg, Pennsylvania, in the Kitzmiller v. The Dover Area School District case. Here Judge John E. Jones III[5], nominated by George W. Bush, made a decisive ruling about Intelligent Design, saying that members of Dover's school lied under oath to hide their religious motivations and that "ID was not science."

I guess you can Google hundreds of references if you want to learn more about the legal proceedings of the case and the insights of the people involved in this trial.

---

5   John E. Jones III - Conservapedia

But the bottom line of the subject then was summarized in the Kitzmiller vs. Dover Memorandum of Opinion (Jones, 2005) published by the United States District Court for the Middle District of Pennsylvania. There, you will find every aspect reviewed in court and its corresponding final ruling.

However, after reviewing most of the available information about this case, it looks that the main line of defense from the Dover Area School District Board to justify the introduction of their proposed hypothesis in the public classrooms was to prove that the ID hypothesis was science, and because it was science, it deserved a place in the student's curriculum alongside the theory of evolution of the species. This, in my opinion, was a very naive strategy to introduce the subject into our public school system, which makes me wonder (due to my lack of knowledge on the topic) how many of the disciplines currently taught in the public school system made it to the classrooms through the justice branch of government. Or, which is the best route with adequate checks and balances for a new body of knowledge like this one to be included as part of the student's science curriculum?

Now I understand what was at stake from both sides of this conflict and why federal judge John E. Jones III decided against the ID movement, saying that the "ID is NOT SCIENCE (NOT SCIENCE, 2005).

IS THE ID HYPOTHESIS SCIENCE?

© 2005 The Patriot-News. All Rights Reserved. Reprinted with permission.

Figure 9 - The Patriot-News, Harrisburg, PA December 21, 2005

Unfortunately, this got the wrong interpretation by the general public, probably due to our limited understanding of the scope of what could be considered to be science when it comes to establishing the boundaries or demarcation of our instinct to inquire and acquire knowledge in disciplines like metaphysics, philosophy, ethics, history, or religion. These disciplines don't rely exclusively on the generally accepted Scientific Method of Analysis based on evidence to explain their empirical observations, which is the case of the ID hypothesis.

For most people saying that a hypothesis is not science gets interpreted as the hypothesis is false, and this interpretation comes from the wrong notion that all science has to be true. But this notion can be quickly challenged by the many efforts to acquire knowledge from past generations well documented in scientific proposals, hypotheses, and theories that were eventually determined to be false and superseded.

However, due to this fundamental misunderstanding, a wrong, probably unintentional message was sent to the general public, saying that because the ID hypothesis was "***not science***," it was false. And that this was the reason why it could not be part of the curriculum of our public school system.

But the real reason that prevented its teaching in our school system was our Constitution's First Amendment Clause (The 1st Amendment of the U.S. Constitution, 2022). This clause guarantees the Freedom of Religion, Speech, Press, Assembly, and Petition and prohibits the government from "establishing or favoring" a particular religion.

This aspect of the ID hypothesis was pronounced during the discovery of evidence in the trial proceedings, where the ID hypothesis's religious/creationist influence/background in its early stages of conception was revealed; this made it very clear to Judge John E. Jones that ID was linked to creationist and religious antecedents, and that the real purpose or motivations behind the school board's efforts to teach the proposed hypothesis in our public school system were to promote religion.

This situation provided an escape to solving the litigation between both parties by leveraging the First Amendment's constitutional provisions establishing the separation between state and church. But unfortunately, at the same time, it created lots of confusion about whether or not the Intelligent Design Hypothesis was or was not science, a subject that, in my opinion, was not adequately argued in court then by its advocates.

So, what am I saying?

I'm saying that I buy into the fact that the ID movement had a creationist/religious background that gave birth to its hypothesis. But if not adequately decoupled from its original roots, it is a valid reason to reject its introduction into our public school system for the constitutional reasons mentioned before.

But I'm not sure that I can buy into the reasons outlined in the **Memorandum of Opinion** of this case to conclude that "*ID is not Science.*"

Let's see why.

4. Whether ID is Science (Jones, 2005)

(1) "*ID violates the centuries-old ground rules of science by invoking and permitting supernatural causation;.*"

Invoking or permitting supernatural causation in any part of an argument is generally considered a **Blind Authority** or **God of the Gasps** logical fallacy. Believers of different religions mostly use this fallacy to explain the cause for some observed natural phenomena that science cannot explain at the time of the argument.

Thus, invoking this fallacy as part of the conclusion of an argument leaves without identifying the cause responsible for the observed effect. Because in most cases, the alluded supernatural cause is not proven to exist, and it is a reasoning error to make a non-existent cause responsible for an existent natural phenomenon.

But the question that we have to ask in this regard is not if the arguer is using an invalid argument to try to prove his point but whether or not using an invalid argument to prove his point has anything to do with the scientific nature of the hypothesis in question.

More specifically, if arguing that human beings and nature are the creation of an intelligent designer (a variation of supernatural causation) means or implies in any way that the ID hypothesis is not science or scientific.

My answer to this question is NO because the Scientific Method of Analysis allows bad arguments to exist as we iterate through its process steps to determine if a given proposition, hypothesis, or theory is true or false.

And the scientific nature of a hypothesis is not determined by whether or not it has bad arguments or logical errors or if it is eventually determined to be true or false. It is determined by its adherence to the methods and principles of science based on evidence to explain its empirical observation(s).

Therefore, constructing a lousy argument with fallacies to try to prove a point does not necessarily mean or prove that a given proposition, hypothesis, or theory is not science or is not adhering to the methods and principles of science based on evidence. It could very well be a valid conclusion with a terrible argument.

(2) *"…the argument of irreducible complexity, central to ID, employs the same flawed and illogical contrived dualism that doomed creation science in the 1980s."*

Like the above, this argument commits the same **Blind Authority** or **God of the Gasps** logical fallacy. And like the previous one, it tells us nothing about the scientific nature of the ID hypothesis since the Scientific Method of Analysis allows for invalid arguments to exist as part of its iterative evaluation process seeking to prove if a given point is true or false.

Therefore, committing this fallacy, as discussed before, does not necessarily mean that the ID hypothesis is not science or scientific.

(3) *"ID's negative attacks on evolution have been refuted by the scientific community. As we will discuss in more detail below, it is additionally important to note that ID has failed to gain acceptance in the scientific community, it has not generated peer-reviewed publications, nor has it been the subject of testing and research."*

There is nothing wrong with the ID's "negative attacks" on evolution and the corresponding refutation from the scientific community. Since this is part of comparing competing hypotheses for a given empirical observation, this is particularly important for hypotheses of past historical events like the ID hypothesis for which no evidence is available or practical experimentation can be carried out to test the predictions of the hypotheses.

But what is relevant here is that the argumentation and refutation process that we go through as part of analyzing competing hypotheses is an integral part of the generally accepted methods and principles of science based on evidence, whose final result should be a true or false determination based on the evidence presented.

Therefore, this should not be a valid reason to conclude that the ID hypothesis is not science or scientific.

On the other hand, I'm not surprised that the ID hypothesis hasn't gained much acceptance in the scientific community and hasn't generated peer-reviewed publications because its advocates have not adequately argued it. The scientific community knows this and is giving its advocates the cold shoulder, sending them a subliminal message to "go back to the drawing board."

Now, scientists can't adequately test all hypotheses of past historical events (like the ID hypothesis) with 100% adherence to the traditional Scientific Method of Analysis because it is virtually impossible today to re-create the experimental conditions of these past events to confirm any predictions. But these hypotheses are very useful to guide future scientific research and make educated guesses about the potential causes of past events to choose the best explanation based on the available evidence.

But this does not mean that these kinds of hypotheses are not science or scientific. If this were the case, for example, disciplines of inquiry of knowledge like archeology, anthropology, economics, artificial intelligence, and social sciences would have to be considered not scientific.

Additionally, the three above-stated reasons in the Memorandum of Opinion of this case concluded that the "ID hypothesis is not science" without considering the difference that exists between our intellectual practice of pursuing the understanding and application of knowledge of our natural and social world, to which we assign the noun science, and the methods and principles of science that we use to acquire this knowledge, to which we assign the adjective scientific.

In summarizing, saying that the *"ID Hypothesis is not Science"* is the same as saying that the ID hypothesis is not pursuing the understanding and application of knowledge of our natural and social world and that it is not adhering to the methods and principles of science based on evidence that we use to acquire knowledge, something that I can't agree with after reviewing their proposed materials.

CHAPTER 5

Is the ID hypothesis scientific?

As discussed before, the word scientific is an adjective. And as an adjective, it is grammatically related to a noun. Its purpose is to expand on the attributes of the noun to describe it better.

And when we say that a hypothesis is scientific, we mean that in addition to being a proposed explanation for empirical observation, it has an additional attribute or characteristic that better describes or complements the nature of the hypothesis by denoting that it has been documented in the pursuit of knowledge and understanding of the world around us, adhering to the principles and methods used in science by following a systematic methodology based on evidence. (The Scientific Method, 2022).

However, scientists from different disciplines of the natural and social sciences use similar but different types of methods to acquire knowledge through observation and experimentation to build their logical arguments and convince others that their hypotheses are accurate descriptions or explanations of our observed reality.

And although their methods may differ in techniques, number of steps, the order in which these steps are executed, and the emphasis that they put on each step, most of them adhere in principle with varying degrees to some or all the following general guidelines:

1. First, we make an observation.
2. Second, we ask ourselves a question about the observation.
3. Third, we research and gather background information regarding the observation and the question we want to answer.
4. Fourth, we form and present a proposed explanation for the question we are trying to answer.
5. Fifth, we make one or various predictions about the observation based on the proposed explanation.
6. Sixth, we test our predictions through experimentation and observation to gather evidence.
7. Seventh, we analyze the evidence obtained to determine if it confirms or denies our prediction(s).
8. Eight, if the evidence confirms the prediction(s), the proposed hypothesis accurately describes our observed reality.
9. Ninth, if the evidence contradicts the prediction(s), the proposed hypothesis is considered a false description of our observed reality.
10. Tenth, we iterate through these process steps as often as we feel necessary to gain enough confidence about the true or false determination of the proposed explanation.

Therefore, if a hypothesis adheres to most of the above basic guidelines, it can be considered scientific. But if the hypothesis deviates significantly from the above guidelines, it is not regarded as scientific.

Notice, however, that the determination of whether or not a hypothesis is true or false is not a requirement of this Scientific Method of Analysis but a conclusion that is obtained as a result of following these guidelines, which means that a hypothesis can be scientific, and be eventually determined to be true or false.

But a hypothesis that is not scientific is virtually impossible to determine whether it is true or false due to the lack of adequate information and evidence to reach this determination.

Does the ID-H meet the SMA criteria?

Now, we can gauge the ID hypothesis's adherence to the above general guidelines to determine if this hypothesis is or isn't scientific.

First, ID advocates have observed the vast diversity and complexity of life and nature on Earth. So the first guideline is met.

And they have generated a fundamental question about the origins of human life and nature. Therefore the second guideline is also met.

They have also researched and gathered lots of background reference information and supporting information from different fields of science to support their hypothesis. So the third guideline is also met.

And they have formulated and presented a proposed explanation for the origins of human beings and nature. Therefore the fourth guideline is also met.

Also, they have generated a series of predictions based on the proposed hypothesis. Therefore the fifth guideline is also met.

And they have proposed several tests for their predictions. Therefore the sixth guideline is also met.

And they have tested their predictions and have presented the evidence obtained. Therefore the seventh guideline has also been met.

They have determined, in their opinion, that the evidence obtained during testing confirms the predictions that they have made. Therefore the eight guidelines have also been met.

And they have determined in their opinion that the evidence obtained during testing does not deny the predictions that they have made. Therefore the ninth guideline has also been met.

And based on the fact that they have published their work, they are probably satisfied with the number of times they have iterated through these process steps to conclude that the ID hypothesis accurately describes our observed reality. Therefore the tenth guideline has also been met.

Hence, we can conclude that the ID hypothesis follows and meets most, if not all, of the basic general guidelines of the Scientific Method of Analysis to be considered a scientific hypothesis.

Now, a few words of caution on the above interpretation of adherence to the Scientific Method of Analysis; the fact that ID advocates follow a systematic method of analysis based on evidence does not guarantee that everyone who reads the presented materials will agree with the information provided, and the conclusions reached. But agreeing or disagreeing with their materials is a different subject than whether or not their hypothesis is or isn't following a systematic method of analysis based on evidence to reach its conclusions.

Some may disagree entirely with all the presented materials, while others will agree. And others may agree or differ to a certain extent in different steps of the process. And this is why peer reviews are so critical within the science landscape to evaluate proposed hypotheses and determine if there is consensus in the methods used, the information provided, and the conclusions reached.

But regardless of whether we all agree or disagree with some or all the information provided, including but not limited to its predictions, proposed tests, the evidence submitted, logical arguments, and conclusions, ID advocates used a systematic analysis approach based on evidence to prove their assertion.

Therefore, the presented ID hypothesis must be considered a scientific one.

Some of you, especially those who disagree with the proposed predictions and test results, may think that the ID hypothesis is not scientific because this evidence is inadequate. But the fact that a hypothesis, as presented, has reasoning errors, false arguments, wrong information, or is eventually determined to be false, does not disqualify it from being considered a scientific one, especially if it adheres to the methods and principles of science based on evidence to explain its empirical observation.

If this were the case, that would undoubtedly undermine the very own nature of science in our search for knowledge.

As a line of reference, think about the many scientific hypotheses and theories that were once considered valid and were later replaced by more accurate explanations of our observed reality for these same reasons.

Yet, there is still this unexplained notion amongst the general public that scientific hypotheses must be true and non-scientific hypotheses must be false. When in practice, the scientific adjective only means that the hypothesis follows a systematic method of analysis based on evidence to explain its observations. And the hypothesis's true or false attribute determination is only achieved after subject matter experts evaluate all documentation submitted to determine if it is or isn't an accurate description of our observed reality. And this is the case because there is no reliable way of knowing upfront if a hypothesis is true or false unless we follow a systematic method of analysis based on evidence to collect the necessary information to reach this determination.

Therefore, the only clear requirement to classify a hypothesis as scientific is that it adheres to a systematic method of analysis based on evidence. And, as long as it adheres to these general guidelines, whether it is eventually determined to be true or false, it has to be considered a scientific hypothesis. And this is the case with the Intelligent Design hypothesis.

The Phlogiston Theory

The Phlogiston theory is an excellent example of a scientific hypothesis that was considered a theory, eventually superseded by other theories[6]. Once tried to explain the combustion and rusting processes, this theory postulated that a fire-like element called Phlogiston was contained within combustible bodies and released during combustion.

(Incredible, I still remember when I took this science class about phlogiston as a kid a long time ago.)

Today it is a superseded scientific theory, which we know is false. Nevertheless, all the scientific efforts and experimentation on the subject back then were well recognized as scientific because they adhered to a scientific method of analysis based on evidence.

Therefore, the Phlogiston theory, although once considered trustworthy and now known as false, is still considered a scientific theory (superseded) because it adhered to these basic guidelines throughout its life cycle. And the same is true for all other superseded scientific

---

6   phlogiston | chemical theory | Britannica

theories that have been replaced for better explanations of our observed reality.

Therefore, adherence to a scientific method of analysis based on evidence is what determines if a given hypothesis is scientific, no matter if;

A. It is proven to be true.
B. It is proven to be false.
C. It is proved to be true and later proved to be false.
D. It is proved to be false and later proved to be true.

And the reason for this is that adhering to a scientific method of analysis based on evidence provides the necessary information and logical framework to make this true or false determination to the best of our knowledge at a given moment.

CHAPTER 6

Inferred ID Argument

I'm sure that there have to be several arguments that can be used to try to prove *that human beings and nature are the results of an act of creation by an intelligent designer and not the product of random natural causes like abiogenesis and evolution.*

However, the most convincing one is to present tangible evidence to prove the existence of this cause (intelligent designer) and to demonstrate that it was capable and responsible for deriving the effect of human beings and nature from the elements of matter at one particular location and moment in time.

But this proposed act of creation happened a long time ago, and there is not enough information on the intelligent designer and the process used by him to carry out the referenced matter-to-life conversion to predict a possible outcome that can be adequately tested.

So, the only realistic option we have from the logical point of view is to use an abductive reasoning mode argument to compare other known causes that could derive the cell's genetic information to choose the best explanation amongst them.

But do we know which empirical observation the ID hypothesis tries to explain? Is it human beings and nature, the carbon-based cell, the genetic information in the cell's DNA, or all of the above?

It is interesting because, throughout the available materials, ID advocates seem to jump from one topic to the other relatively quickly without providing enough information or evidence of the necessary sequence of events to explain each one of these observations.

And this is where spelling out a clear argument with the empirical observation being explained helps reduce the amount of speculation about what is being proposed. But with no clearly stated argument, I have no option other than making some "educated" guesses to come up with my interpretation of what I perceive is the empirical observation being explained and the type of argument being presented.

First, I assumed that the empirical observation used by the ID advocates as the core of its argument is the Functionally Specified Information (genetic information) contained in the DNA of all carbon-based living entities. I make this assumption based on the numerous times that this genetic information is referenced throughout the ID materials to argue the statistical impossibility of other natural causes like evolution and abiogenesis to produce this information at random.

Second, I also assumed that this information is used in connection with the intelligence attribute of our species to infer a past intelligent designer cause.

Third, although evolution is also extensively referenced throughout the ID materials, and like abiogenesis is a random natural process that affects carbon-based living entities, I ruled it out from the inferred ID argument because, as of this day, there is no evidence to prove that the evolution theory can explain the origin of carbon-based living entities, its cells, or its genetic information.

Fourth, I kept the abiogenesis hypotheses as part of the inferred argument because these are <u>potential explanations</u> parallel to the line of action of the proposed intelligent designer that could eventually explain a matter-to-life conversion process for the emergence of the first carbon-based cells.

Therefore, on the assumption that the origin of human beings and nature on planet Earth was an isolated historical event that happened a long time ago for which there is no available evidence of the cause that derived its existence, it seems evident that there is no other option other than to create an abductive reasoning mode argument with premises related to our intelligence attribute and the information contained in the cell's DNA of carbon-based entities, to infer an Intelligent Designer as a possible cause for the existence of human beings and nature.

However, remember that the abductive reasoning mode argument is primarily used to formulate proposals of probable causes that don't necessarily follow from its premises. And because of this, it has some inherent limitations regarding the certainty of the conclusion it reaches when compared to deductive or inductive reasoning modes arguments.

And this is the case because abduction reasons by comparing an observation to a specific rule, fact, or another observation(s) to make an educated guess about the possible cause for the observation. And it produces an explanation of how two things can be possibly connected, usually referred to as an inference to the best explanation or as a "speculative hypothesis" that can work as a starting point for further investigation.

However, in this abductive reasoning argument, the best explanation of a series of competing explanations is only the best explanation. And no matter how good this explanation is, it will not produce the necessary evidence to prove 100% the proposed cause for the explained empirical observation. And its confidence level as the best explanation is highly dependent on the comparison method used to reach its conclusion.

Hence after reviewing most of the available ID materials seems to me a reasonable inference that ID advocates choose to argue their hypothesis about the creation of human beings and nature by comparing it to existing abiogenesis hypotheses (Hypotheses about the origins of life, 2016), trying to establish a logical connection between the genetic information contained in our cell's DNA, and the intelligence attribute of human beings.

They established as the **RULE** of their argument that the type of genetic information contained in our cell's DNA can only be the product of the mind of an intelligent designer.

And they presented as the **EFFECT** of their argument that existing abiogenesis hypotheses have not been able to successfully generate this type of genetic information through random chemical/physical processes.

To conclude, that the possible **CAUSE** of the existence of human beings and nature can't be a random chemical/physical process like abiogenesis but the intentional creation of an intelligent designer.

Therefore, I will venture to say that the following abductive reasoning mode argument is an adequate one for what ID advocates are trying to prove:

Inferred Intelligent Design Argument

First Premise (RULE)
Only intelligent designers are capable of generating large amounts of information.

Second Premise (EFFECT)
Known abiogenesis hypotheses can't successfully demonstrate how large amounts of information can arise from inert matter through purely random chemical processes.

Conclusion (CAUSE)
Therefore, human beings and nature are the results of the act of creation of an intelligent designer.

This argument can be worded in half a dozen ways, as in the previous statements. Still, the core of its premises about intelligence, genetic information, and the conclusion it is trying to reach is not far from the above.

Again, I may be wrong in my interpretation of some of the available information on this subject, but for sure not too far from the kind of argument that ID advocates have presented up to this day.

To prove this argument valid, we must do a detailed logical analysis of its premises and their relationship to its conclusion.

CHAPTER 7

Is the Inferred ID Argument Valid?

As a guideline, generally speaking, a valid argument leads you to an accurate conclusion making use of sound reasoning logic with true and unambiguous premises containing the evidence (reasons) for which there is no possible situation in which the conclusion can be false. And any false premise or any illogical relationship of a premise to the conclusion invalidates the argument.

However, remember that in science, we may not know today whether a given argument is absolutely valid since future acquired knowledge and new scientific discoveries may eventually challenge this conclusion. But we can always test any argument at any time by analyzing its premises and the existing relationship between these and its conclusion to determine if it is a valid logical one.

To do this, we must determine if its premises are true or false and if they maintain a good logical relationship between them and the conclusion.

First Premise (Rule)
Only intelligent designers are capable of generating large amounts of information.

True or False?
Generally speaking, this is a true premise, particularly when we can recognize in the physical media where the information is contained, large amounts of complex/organized logical patterns or structures that can be associated as being the product of the mind of an intelligent entity, which is the case for example, of the genetic information contained in our cell's DNA.

Relationship to the Second Premise
What kind of relationship exists between this first and second premise of the inferred ID argument?

In other words, what kind of relationship exists between the fact that only intelligent designers can generate large amounts of information and the statement that known abiogenesis hypotheses have not demonstrated a process capable of generating this large amount of information at random from inert matter?

The implicit relationship is that the intelligent designer and the abiogenesis hypotheses are both potential causes that could derive from the existing matter the genetic information contained in the cell's DNA. In other words, both causes are assumed to be competing explanations for the presence of this information in the cell's DNA.

However, we know that abiogenesis hypotheses are trying to demonstrate a process by which organic compounds and simple life forms could have emerged randomly from inert matter under the proper environmental conditions. And we also know that until now, these hypotheses are not trying to explain or demonstrate a process capable of producing the genetic information in the cell DNA of complex multicellular organisms.

So, there seems to be a significant gap between the effect that an intelligent designer could derive from the existing matter to create this type of information and the effect that the abiogenesis hypothesis could derive from the existing matter to make organic compounds and simple cell organisms.

In other words, at this point, abiogenesis hypotheses are trying to find a random environmental/chemical process capable of producing organic molecules and single cells organisms like prokaryotes (Cyanobacteria, Archaea), protocells, etc., and not concerned about what process could derive the genetic information contained in the cell's DNA of carbon-based living entities like Eukaryotes (plants, animals, fungi, protist...etc.)

These are two different types of hypotheses pursuing the explanation of two different events, possibly in the same line of actions necessary to explain the existence of human beings and nature.

Relationship to the Conclusion
What kind of relationship exists between this first premise and the conclusion of the ID argument?

Indeed, having these large amounts of information in our cell's DNA suggests that this recognized logical pattern of four letters (ATGC) of information can be the product of an intelligent mind.

But is the fact that we possess this information in our cell's DNA enough evidence to prove that an intelligent designer created human beings and nature on planet Earth?

Could other causes also explain the presence of this complex information in our cell's DNA that does not require an intentional creation process on planet Earth at the hands of an intelligent designer?

Certainly, both sides of this coin are possible, and we will review this in more detail in the Analysis's Comparing Competing Hypotheses section.

Second Premise (Effect)

Known abiogenesis hypotheses can't successfully demonstrate how large amounts of information can arise from inert matter through a purely random chemical process.

True or False?

The above premise establishes as the effect of the ID argument that abiogenesis hypotheses have not demonstrated a process capable of creating large amounts of information at random from inert matter.

And although, at first glance, this seems to be a valid premise, further examination demonstrates that it is a misleading premise that fails to provide adequate evidence to support the conclusion of the inferred ID argument.

The reason that I say this is because, to the best of my knowledge up to this day, no currently known abiogenesis hypothesis is trying to demonstrate a process on how to generate large amounts of information like the one that is contained in our cell's DNA, at random from the inert matter.

Suppose I got this right from my research notes. In that case, all currently known abiogenesis hypotheses are trying to explain the origin of the carbon-based cell on Earth by attempting to replicate past environmental conditions by which simple life forms (no large amounts of genetic information) emerged from inert matter in a purely random manner. And no single effort is being made by these hypotheses' proponents to demonstrate a process that can explain the cause responsible for this DNA information in the cells.

Therefore, from the logical point of view, abiogenesis hypotheses can't fail to demonstrate a process that they aren't pursuing to demonstrate. Hence it is not true that they have failed in this endeavor, as well as it is not true that they will ever succeed either since they are not attempting to do so.

Therefore, this second premise is misleading and does not provide any supporting evidence to the argument's conclusion; hence, it is considered false.

Relationship to the First Premise
This relationship was covered in the analysis of the first premise.

Relationship to the Conclusion
Now, suppose any of these abiogenesis hypotheses succeeds in proving their point. In that case, this will only prove that it is possible to generate simple life forms (not large amounts of information) from nonliving matter through a random chemical and physical process. And, if all of these abiogenesis hypotheses continue to fail to prove their point, this will only prove that we still don't know a method capable of generating simple life forms from nonliving matter through a random chemical and physical process. Still, neither of these outcomes will produce evidence to support that abiogenesis hypotheses are capable of deriving the genetic information in our cell's DNA from nonliving matter.

Therefore, whether or not abiogenesis hypotheses succeed or fail to generate simple life forms from inert matter, it will not provide any supporting evidence or valuable information to prove what cause was responsible for the creation of human beings and nature.

In other words, there is no logical relationship between the failure or success of abiogenesis hypotheses to create simple organic compounds and life forms from nonliving matter at random and the conclusion that human beings and nature are the act of creation of an intelligent designer.

This type of reasoning is a well-known **Does Not Follow** (Non-Sequitur) logical fallacy Where the conclusion of the argument does not follow from one of its premises, or the reason or evidence presented in the premise is irrelevant to the conclusion of the argument, causing people to spread inaccurate information by making unsupported claims from the inadequate relationship of one thought to another.

Conclusion (Cause)
Therefore, human beings and nature are the results of the act of creation of an intelligent designer.

True or False?
It is evident from the analysis of the argument's premises that the conclusion of the inferred ID argument can't be adequately supported with a first premise that fails to establish a good relationship with the second premise, as well as with a false second premise that commits a logical reasoning error (non-sequitur) and lacks adequate evidence to support its conclusion.

Therefore concluding that human beings and nature are the creation of an intelligent designer based primarily on the false assumption and the lack of evidence from abiogenesis hypotheses to generate large amounts of information is considered to be an **Argument from Ignorance** fallacy because abiogenesis hypotheses at this point are not pursuing an explanation of the cause responsible for the creation of this information.

In addition, claiming the intervention of a divine entity or a variation of it without any evidence to support its existence (intelligent designer) to account for some natural phenomena that science cannot explain at the time of the argument is known as the **Divine Fallacy** or **God of the Gaps** fallacy.

Therefore, working on the assumption that I have made the correct inference about the ID hypothesis argument, the above analysis proves that ID advocates could not put together a convincing argument to prove that their hypothesis accurately describes our observed reality.

This situation was recognized by Stephen C. Meyer in his book Signature in the Cell when he wrote;

"Now I repeatedly found myself in the position of having to defend an argument in sound bites that my audience did not know well enough to evaluate. How could they? Perhaps the central argument for intelligent design, the one that first induced me to consider the hypothesis, had not been explained adequately to a general, scientifically literate audience."[7]

And suppose ID advocates want to prove that their proposed hypothesis is true. In that case, additional iterations through the scientific method of analysis will be required in search of new evidence to build a different argument that can support their conclusion.

---

7   Meyer, Stephen C. Signature in the Cell (p. 5). HarperCollins. Kindle Edition.

CHAPTER 8

Is the ID Hypothesis False?

Does the fact that the inferred ID argument is not valid prove that its conclusion is false? In other words, is it false that human beings and nature are the creation of an intelligent designer?

The answer to this question is **NO!**

Just because the inferred ID argument commits reasoning errors or logical fallacies does not prove false the possibility that an intelligent cause or entity could have been responsible for the creation of human beings and nature. Because this could very well result from a poorly constructed argument trying to prove a valid point. And assuming that the conclusion of an argument is false because its premises do not support the conclusion properly is considered a **Bad Reasoning Fallacy**.

The only thing we can conclude so far is that with the information and the inferred argument presented, it is impossible to prove that human beings and nature are the creation of an intelligent entity.

CHAPTER 9

In Search of the Truth

On the assumption that I have made the correct inference about the ID hypothesis and its argument, the previous analysis shows that ID advocates were not able to put together a convincing, sound, logical argument to prove that their hypothesis is accurate, correct, and verifiable by facts of our observed reality.

But unfortunately, the original question about the origins of human beings and nature remains unanswered.

Is it true that we are the creation of an intelligent entity? Or is it possible that our existence here on planet Earth results from other natural causes?

Indeed, these are fundamental questions that continue to elude the human intellect. But we will never be close to finding any answers if we ever quit trying to find the truth.

Therefore, I will attempt to double-click one more level below the surface of the assumptions that are the foundation of the ID hypothesis, hoping to find a different point of view that can help us move this paradigm one step forward.

As most of us know, the scientific method of analysis is the most widely used and trusted guideline of science to determine if a proposed explanation of our observed reality is true or false.

This method requires, among other things, that we derive predictions from the proposed explanation about the results of some future experiments to perform these experiments and gather the necessary evidence to be analyzed. Then if the evidence obtained agrees with the predictions derived from the explanation, the hypothesis is considered true. But suppose the evidence obtained contradicts the predictions derived from the explanation. In that case, the proposed explanation is deemed false, pushing the timetable back to the drawing board to review and re-think the proposed explanation and its predictions.

The good news is that although this scientific method of analysis at first glance imposes a strict discipline to be followed, it should be a straightforward process to execute if you are familiar with the proposed subject matter. And it should help refine our approach and filter unnecessary information to improve the message being communicated.

But the bad news is that even if you follow this widely accepted method of analysis, there is no guarantee that everyone who reviews your proposed hypothesis will agree with your empirical observation, its explanation, the tests performed, the measurement methods employed, or the interpretation of the results obtained during testing because this is the intrinsic nature of science at its best.

And although this method seems to be a simple approach to making true or false determinations, sometimes applying it to proposed hypotheses is not straightforward, especially when it comes to fields of science that use different inquiry methods to acquire knowledge, with different criteria to gauge how accurate, correct, and verifiable a given empirical observation is with regards to known facts, and available evidence.

Therefore, we must be open to the fact that not all proposed hypotheses or explanations use the same scientific method of inquiry to acquire knowledge nor the same type of reasoning mode argument to prove their points.

And, because evidence can be anything that is presented in support of a proposed prediction, this support can range from very strong to very weak depending on the tests performed, the measurement methods employed for the variables of interest, the actual results obtained during experimentation, and the final interpretation of these results.

Therefore, not all the evidence presented on the proposed hypotheses will provide the same confidence level to support their conclusions.

As a result, depending on the field of science and the type of argument used, some hypotheses bring strong evidence that provides direct proof of the truth that effectively rules out other competing explanations for the same empirical observation. In contrast, others get weak evidence that is merely consistent with the proposed explanations but doesn't do a good job ruling out other possible explanations.

This becomes particularly true for hypotheses about causes of historical/past events that happened a long time ago, for which there is no direct proof at present of the responsible cause of the empirical observation, and for which no experimental test can be carried out today with a desired future outcome that can be compared to the predictions of the hypothesis.

But generally speaking, the more everyone agrees with the information presented in the materials of a hypothesis, and the fewer conflicting opinions, the more likely it is that the proposed hypothesis will be considered an accurate description of our observed reality.

And the more that everyone disagrees with the information presented in the materials of a hypothesis and the more conflicting opinions about what is being proposed, the more likely it is that the proposed hypothesis will eventually be considered an inaccurate description of our observed reality.

But even in the above two scenarios, due to the ever-changing nature of our acquired knowledge, a hypothesis considered to be true (theory) might not be true forever. And a hypothesis deemed to be false could not be false forever either.

So, can we find the truth about any given observation in our observed reality?

Well, in the immense scale of our space/time dimension as we move forward, maybe temporarily.

CHAPTER 10

Testing a Theory False

If there is something that science can be credited for in its relatively short life span, it has helped humanity better understand the reality in which we all live.

Science has given us the tools to observe the environment around us to help us explain how things work to predict future behaviors that can be applied to solve the challenges that we face for the survival of our species.

And this has been possible to a great extent because science, at its most fundamental level, operates based on the principle of non-contradiction. This means it does not simultaneously allow a statement to be true and false except for the new field of Paraconsistent logic (Graham, Koji, & Zach, 2022). Because allowing contradictory statements to coexist in science inhibits us from proving or disproving anything, and it would never be possible to tell the difference between a truth and a lie.

However, even with the most significant achievements we claim through our scientific efforts, we must admit that science has inherent limitations that prevent us from finding the absolute long-term truth of our perceived reality. And we are probably the most influential factor in this effect. Because we are physically and intellectually limited in many ways by our capacity to perceive, interpret and interact with our perceived reality.

And there is no other place where this limitation is more evident than in all of our scientific theories by the nature of the results we obtain during experimentation. Because all of our theories have a limited scope of validity, which is determined by the boundaries of our capabilities to observe, our capacity to imagine possible causes, the design of our proposed experiments to test predictions, the measuring methods that we use to collect test results, as well as our intellectual capacity to interpret the evidence and reach logical conclusions that can be applied to solve our day-to-day problems.

Therefore, although our scientific truths are good enough for all practical purposes in short to midterm, we can only consider them incomplete, partial truths in the long term.

The most we can do is to demonstrate that a theory is valid within the boundaries of the body of evidence collected during its test and experimentation. But there is always a possibility that new future observations or further discoveries of evidence could bring to light a new set of boundaries that can prove it false.

This was the case, for example, of the traditional Newton's Laws of Physics (Newton's Laws of Motion), which were considered for a long time to be the only explanation of the physics involved in objects moving in our observed reality. But as we moved into the scale of atomic and subatomic particles, this explanation failed to describe their observed behaviors adequately. It was eventually replaced by Quantum Theory (Quantum Theory, 2022) explanations.

Nevertheless, Newton's Laws of Physics are still valid within the boundaries of the body of evidence presented initially by Newton at what we consider the macro scale of matter. But not right within the limits of the body of evidence presented by Max Planck, Albert Einstein, Louis de Broglie, and Heisenberg Werner at what we consider the nanoscale of matter.

So, in science, it is generally accepted that evidence can never prove a theory to be true forever, but it can certainly prove it false in a split second.

And suppose you can prove any theory to be false. In that case, chances are that you are using an excellent scientific methodology to raise the necessary evidence to establish facts inconsistent with the previous body of evidence obtained during the initial experimentation results.

And this is how science makes its most significant contributions to humanity by proving wrong established theories and replacing them with new ones.

Therefore, as a result of this, the most significant contribution that any given hypothesis or theory can make to advance science is to eventually fail to describe our reality accurately, to be wrong because, upon its failure, a new and improved description of our perceived reality is born in the form of a new theory as our next best approximation of scientific truth.

CHAPTER 11

Falsifying a Hypothesis

But can this useful concept of proving false scientific theories to advance science be applied to hypotheses as a quick go, no-go test to determine if a hypothesis is scientific as it was proposed by the British philosopher of science Karl Popper (Mcleod, 2020) in 1983?

Well, some definitions of Karl Popper's falsifiability must be clearly understood.

The first one is what it means to say a hypothesis is scientific.

As discussed before, the descriptive attribute or adjective "scientific" is assigned to a hypothesis if the hypothesis being studied adheres to or is characterized by the methods and principles of science, following a systematic methodology based on evidence.

Therefore, by definition, a hypothesis that does not adhere to or is not characterized by the methods and principles of science following a systematic methodology based on evidence is not scientific.

And the second one is, what does it mean when we say a hypothesis is false?

As discussed before, the descriptive attribute or adjective "false" is assigned to a hypothesis when the hypothesis predictions are tested and the evidence obtained from the test results is found to contradict its predictions. Therefore, by definition, a hypothesis whose test results are found to contradict its predictions is considered to be a false hypothesis.

Now, the question that needs to be answered is if it is possible to determine the scientific nature of a hypothesis by conceiving a test (not executing it) that could yield a result to prove a hypothesis to be false (falsify it), as proposed by Karl Popper.

Knowing the relationship of these attributes as they relate to hypotheses, we can use this as a guideline to test Popper's falsifiability test.

First, note that it is evident that both of these attributes describe different aspects or characteristics of a hypothesis.

The first (being scientific) relates to the adherence of a hypothesis to the methods and principles of science based on evidence used to study a proposed explanation of an empirical observation.

And the second one (being false) relates to the determination reached when the experimental test results of the hypothesis contradict its predictions.

But within the framework of science, it is generally accepted that it is first necessary to follow the methods and principles based on evidence to create a scientific hypothesis that can then be tested to determine if it's a true or false description of our observed reality. And this is the case because adhering to these scientific methods and principles generates the evidence that makes this true or false determination possible.

And we know this by experience because we know many true scientific hypotheses and many scientific hypotheses that are false. But all of them, regardless of whether or not they were considered true or false at a given time, are scientific because they followed the methods and principles of science based on evidence to present their arguments and facilitate this true or false determination.

While on the other hand, we also know many "non-scientific" hypotheses that lack adequate research information, testable predictions, and evidence that are difficult to classify as either true or false.

Therefore, as a general guideline resulting from the existing relationship that exists between these two attributes as they relate to hypotheses, it is a requirement for a hypothesis to be scientific first before it can be determined to be true or false because adherence to the methods and principles of science based on evidence provides the necessary information to make possible this determination.

Testing Karl Popper's Falsifiability Concept

Now, knowing the relationship that exists between being scientific or nonscientific and being true or false as it relates to hypotheses, we can use this knowledge to test Karl Popper's falsifiability concept.

In essence, Karl Popper's Falsifiability concept states that a hypothesis is scientific if it is possible "to conceive" a test that could give us a result to prove the hypothesis to be false. Or a hypothesis is nonscientific if it is not possible "to conceive" a test that could give us a result to prove the hypothesis to be false.

But when you search the available literature for examples of how this falsifiability concept is applied to hypotheses to determine its scientific nature, most authors use simple statements about empirical observations that lack the necessary elements that characterize hypotheses to illustrate the concept.

And although it is possible to summarize the conclusion of a hypothesis' argument in a simple statement or to consider any prediction statement of a proposed test outcome to be a hypothesis, not all simple statements contain the elements required to be considered as a hypothesis (Helmenstine, What are the elements of a good hypothesis?, 2019). And using these simple statements as examples to explain how to apply Popper's falsifiability concept creates confusion.

I bring this into consideration because hypotheses are tentative explanations about the causes of observations of the natural world that are appropriately documented following the generally accepted guidelines of the Scientific Method of Analysis (Conley, 2022), which are used as the basis for experimentation to determine if the given explanation is a true or false description of our observed reality.

Hypotheses usually provide enough information to understand the proposed relationship between known independent variables and an unknown dependent variable's expected outcome (or prediction).

And they are often summarized following a logical IF, THEN relationship.

For example;

If I don't drink *coffee* before 10:00 am,
Then I will get a *headache* by noon.

Notice that my headache depends on whether or not I drink my coffee (independent variable) every day before 10:00 am. And this proposed explanation for my noon headaches can be tested and validated as confirmed every time I don't drink coffee in the morning and get a headache by noon.

These essential elements and relationships between independent and dependent variables of typical hypotheses are missing in almost all examples of general observations used to illustrate this falsifiability concept.

For example, "***All swans are white.***"

The previous statement is an excellent statement about a general observation that could be used to trigger discussions on the white swan's topic. But without any other information at hand, it is virtually impossible to determine if it is related to any documented hypothesis or even guess any independent or dependent variables and much less if it is a true or a false statement.

To do so, someone must investigate if the topic has been previously researched and what information is available to support it. Or someone will have to assess all available swans over the face of planet Earth (if possible) to confirm that all of them are white.

Or, someone will have to find at least a black swan (in the southeast and southwest regions of Australia) to prove that black swans exist and that not all are white.

But one way or the other, to determine if the "*All swans are white.*" statement is a hypothesis and whether or not it is a true statement, more information than just a simple unsupported claim or statement is required. And this is where adhering to the scientific method of analysis to properly document a hypothesis on the subject helps.

Otherwise, every statement about our everyday observations, like the one above, could be classified as a scientific hypothesis, which doesn't seem to align with a hypothesis's generally accepted definition.

Now, the "*All swans are white*" statement could very well be the conclusion of a well-documented scientific hypothesis that followed the methods and principles of science based on evidence to prove that all existing swans over the face of the Earth are white. But used as a stand-alone statement without the supporting elements of a specific hypothesis is not a good example to illustrate Popper's falsifiability concept since you are not referring to a hypothesis but a simple statement or observation.

Hence, the potential result of using this approach is that we can end up classifying many simple statements about empirical observations as scientific hypotheses because there is always a good possibility that we can "conceive" a test result like;

"If I can find a black swan, this statement can be falsified or proved wrong."

This will lead you to conclude, as per Popper's falsifiability concept, that the "All swans are white." statement is a scientific hypothesis when it does not even meet the basic requirements of a hypothesis and ignores some of the most fundamental principles of science including the meaning of the scientific attribute or adjective.

Therefore, the claim that the "All swans are white." statement is a scientific hypothesis based on a conceived falsifiability test with no evidence of the existence of a black swan is flawed reasoning, especially when there is no supporting documentation to demonstrate that the statement is adhering to the methods and principles of science based on evidence to explain its observation.

Now, you can prove any simple statement and any complex hypothesis false by getting evidence contradicting their main argument's premises or conclusion.

For example, I can prove that a well-documented hypothesis and a simple statement like "*All swans are white.*" is false if I can find a black swan. But **only if I find a black swan** to demonstrate with objective evidence that not all swans are white.

However, Popper's falsifiability concept is not about the common practice of the science of proving hypotheses or statements to be false by contradicting their premises or arguments with real objective evidence.

Popper's falsifiability concept is primarily a hypothetical/intellectual exercise created to make a relatively quick determination of the scientific nature of a hypothesis, which does not necessarily require that the conceived falsifiability test is carried out to obtain tangible evidence of the existence of the false test result. And, if you can conceive such a test result, as per Popper's falsifiability concept, this is enough "evidence" to allow you to make an "extrapolation" to conclude that the given observation or statement is a scientific hypothesis, an approach that in my opinion is a long stretch with several logical inconsistencies.

First, the hypothetical false test result is not confirmed to exist since there is no requirement to carry out the conceived falsifiability test to prove its existence. This contradicts the generally accepted practice in the science of gathering evidence to determine if the test results contradict the hypothesis prediction's before it can be determined that it is false.

Second, this concept does not establish the logical connection between this proposed/unconfirmed false test result and the adherence of the proposed hypothesis (or observation) to the methods and principles of science based on evidence.

Third, the scientific nature of the hypothesis (or observation) is determined without even knowing if the hypothesis is appropriately documented and is adhering to any methods or principles of science based on evidence to explain its empirical observation.

Fourth, this concept "flows" opposite to the established logical relationship between these two attributes relating to hypotheses. Which first requires a hypothesis to be scientific (adhere to the methods and principles of science) to provide the necessary supporting documentation for evaluation before it can be determined to be false. And not the other way around by first determining that it is false and concluding that it is scientific without any supporting evidence or documentation.

Therefore, we can conclude that a hypothetical test that has not been carried out and that has not produced any supporting evidence or results can't prove any hypothesis to be false. And much less lead us to infer that the hypothesis is following any methods or principles of science based on evidence.

Then, I pause and wonder for a moment. What kind of purpose or real-life application does this proposed go, no-go falsifiability test has for science? Does it make sense to use this concept to argue about the scientific nature of a hypothesis?

Indeed, as with any other argument with unconfirmed or false premises, the conclusion about the scientific nature of a hypothesis can't be sustained through the use of this reasoning logic.

Now someone can claim that if we know as a fact that a hypothesis is false, it is because we probably reached this conclusion by reviewing all the necessary information that is required to make this determination, which confirms that some scientific methods based on evidence were used to explain its empirical observation. Therefore it can be considered to be scientific.

And this is a valid argument as long as the false determination is reached after reviewing the necessary information to get this determination.

But there is a big difference between these two approaches because to know as a fact that a given hypothesis is false requires at least the evaluation of some documentation that uses a methodology based on evidence.

But conceiving an unconfirmed hypothetical test result that could prove a hypothesis about an empirical observation to be false does not produce any documentation or evidence to prove the scientific nature of a proposed statement or a hypothesis.

Therefore, it is a reasoning error to conclude that a conceived falsifiability test with no practical test results is adequate evidence to prove the scientific nature of a hypothesis (or an observation).

So, maybe it is about time we leave some of these old concepts and practices out of our day-to-day scientific propositions and use this time in search of evidence to support our arguments.

Blue Winged Unicorns

To illustrate how the use of the falsifiability concept can lead us to incorrect conclusions about the scientific nature of straightforward claims, let's review a claim or statement that can be falsified in many ways that do not follow an evidence-based method or principle of science to explain its "empirical observation."

But before we get too far into it, let's review the notions of existence and evidence to help us understand the exercise.

EVIDENCE
Evidence is the foundation of all humankind's knowledge which makes possible the current state of technologies that has been developed up to this day in all fields of science, and it is the irrefutable proof and the most robust criteria that we have to determine whether a given belief or proposition is true or false.

And thanks to the effectiveness of the principles and methods of science, knowledge is no longer hostage to traditions or dogmas but based on the observations of nature supported by objective evidence, making it very difficult for anyone to create false claims without providing adequate evidence to support them.

EXISTENCE

On the other hand, the notion of existence refers to the act of actually being there and being able to interact through our physical senses with all that is real, as opposed to that which is merely imaginary within the realms of our mind.

Everything that exists is observable (including but not limited to our vision) and has direct or indirect evidence of its existence from the measurable effects they produce on other things.

But if there is no observable evidence of its existence, there is no way we can know about them. If it is unobservable, how should we know that something exists? If it is not triggering your biological senses or having some observable effect on other things around us, how can we know where it is, what it is, or what it looks like?

Therefore, claiming the existence of something without presenting direct or indirect evidence of its existence is a contradiction and a reasoning error because there is no way to confirm this claim as true.

Now, a friend of mine makes the following statement or claim: "Blue-winged unicorns exist."

My first "instinctive reaction" to this claim is that it is untrue because I have never seen one personally.

But the fact that I have never seen one is not absolute proof of their non-existence. Maybe they do exist somewhere out there, but I have never been lucky enough to see one of them in the same place at the same time.

So, from a logical point of view, I'm probably biased based on my previous experiences. But I have to recognize that there is a possibility that a beautiful Blue Winged Unicorn is furrowing the skies somewhere out there. As well as, there is a possibility that there is none of them in existence.

Although it is generally accepted that the burden of proof of any claim falls on the hands of the person making a claim, my dear friend has not been able to show me any evidence of the existence of his Blue Winged Unicorn.

Hence, I decided to propose several tests to see if we can gather any direct or indirect evidence of the existence or non-existence of these beautiful animals in our observed reality and use these as the baseline to test Popper's falsifiability concept.

As per Popper's falsifiability concept, if I can conceive a test whose result could prove this claim to be false, the claim is falsifiable and scientific.

Conceived Falsifiability Test # 1 - Visual Inspection
If I look for the blue-winged unicorn and can't see it, I could prove this claim false.

Me: OK. Show me, where is the blue-winged unicorn?

My Friend: It is beside the big tree in the backyard. But I forgot to tell you that blue-winged unicorns are invisible.

Conceived Falsifiability Test # 2 - Detection of Footprints
If I pour white flour on the floor where the blue-winged unicorn walks and I can't see any footprints on the ground, I could prove this claim false.

Me: Let's spread some flour around the tree to try to capture the unicorn's footprints.

My Friend: That will not work because Blue Winged Unicorn floats in the air.

Conceived Falsifiability Test # 3 - Sound Recorder
If I install a sound recorder near the area where the blue-winged unicorn is, and I'm unable to capture the sounds he makes when he eats, breaths, walks, flies, etc., then I could prove this claim to be false.

Me: Let's install a sound recorder near the tree area where the blue-winged unicorn is to capture the sounds that he makes.

My Friend: Well, you will not be able to hear it because blue-winged unicorns do not emit any sounds or noise at frequencies within the human beings' audible range.

Conceived Falsifiability Test # 4 - Infrared Signature
If I use an infrared camera to detect his body heat and not get any infrared signature in the image, I could prove this claim false.

Me: Let me use an infrared camera to detect body heat.

My Friend: Well, that will not work either because blue-winged unicorns don't have a visible heat signature in the infrared spectrum of light.

And we keep going down this road to find out that every proposed test I can conceive to prove his claim false has a particular reason why it will not work.

Now, what I have found in the available literature about the falsifiability concept is that its advocates interpret the answers from my friend to the above questions as "evidence" that the blue-winged unicorn's statement or "hypothesis" is not falsifiable or that it can't be proved to be false just because my friend is not willing to run these tests. And I'm afraid I have to disagree with that interpretation because without running the proposed tests and analyzing their results, there is no evidence to reach that conclusion.

I think that the "Blue Winged Unicorn Exists." statement or "hypothesis" can be proved to be false(falsified) by running any one of the above-proposed tests. If we can't get any direct or indirect evidence of its existence, then this lack of evidence proves that blue-winged unicorns do not exist.

Those 360 degrees pictures or videos and IR files around the tree and the floor where the blue-winged unicorn is supposed to be are evidence of absence (DeMichelle, 2021) that the blue-winged unicorn is missing or that it does not exist where my friend claims it to be. And the fact that we can't see it detect its footprints on the floor or record any sounds or infrared signatures proves its non-existence.

My friend makes all of those excuses about the blue-winged unicorn being invisible, floating in the air, making noises outside human's audible range, and having heat signatures outside the infrared spectrum of light, precisely because he fears that if I run these tests in the area where he claims that the blue-winged unicorn is, I will not get any evidence of its existence, but a lot of proof of its non-existence.

Therefore, if I were to apply the falsifiability concept to the claim of the existence of the blue-winged unicorns, I would have to conclude that this claim can be falsified (proved to be false) by gathering the corresponding evidence of the non-existence of the blue-winged unicorn in the vicinity of the big tree in the backyard, which will lead us to conclude as per Popper's falsifiability concept that my friend's claim is scientific.

However, as you may already realize, this claim is not scientific since it is not supported by any principle or method of science based on evidence.

Now, philosophers of science are still debating whether or not the absence of evidence is the same as evidence of absence and what, if anything, should constitute acceptable evidence in both cases. A big part of the debate concerns the long-standing question of science about whether or not it is possible to prove a negative or the non-existence of something. As we will see later, this debate is pivoting around the lack of an adequate definition of the concept of existence in our observed reality.

However, everyone seems to agree that it is possible to prove the non-existence of something as long as we can make the observation and the physical boundaries of the questioned existence of that something are well defined. For example, it is possible to prove the non-existence of a "pink crawling monkey" in the basement of your home. You only need to look in the basement to determine if it is or is not there.

And this is why I choose for this illustration the location of the existence of blue-winged unicorns to be around the big tree in the backyard.

In conclusion, the fact that you can falsify a given statement or "hypothesis," as per Karl Popper's falsifiability concept, is not evidence that such a statement or "hypothesis" follows a scientific method or principle of science based on evidence to prove its claim.

Truliviability Concept

For illustration purposes, on a parallel path to Popper's falsifiability concept, let me introduce the concept of Truliviability. This concept states that a hypothesis is scientific if it is possible "to conceive" a test that, amongst its results, there is one that can prove the hypothesis to be true.

Or a hypothesis is non-scientific if it is not possible "to conceive" a test that, amongst its results, there is one that can prove the hypothesis to be true.

Now copy and paste all of the Testing Karl's Popper Falsifiability Concept paragraphs here, and replace falsifiability with truliviability, and false with true and vice-versa.

You will soon realize that neither one of these two concepts (truliviability nor falsifiability) can effectively argue the scientific nature of a hypothesis with a premise that lacks objective evidence and has not been confirmed to be true or false and without verifying that the proposed hypothesis is documented through the use of an evidence-based scientific method.

Because "conceiving" a hypothetical test that amongst its results there is one that can prove the hypothesis to be true or false does not produce any evidence to reach either of these conclusions and much less infer the hypothesis' adherence to the methods and principles of science based on evidence.

And as we all know, any conclusion of an argument that relies on a false or unconfirmed premise can't be considered trustworthy.

Is the ID hypothesis falsifiable?

ID advocates developed several hypothetical tests to prove their hypothesis was falsifiable and hence scientific. But what does this mean for the Intelligent Design hypothesis or any hypothesis? As discussed before, their hypothesis is scientific not because we can come up with many hypothetical test results to prove that it can be falsified but because its advocates used a scientific analysis method based on evidence to build the argument for its conclusion.

And it is scientific regardless of whether or not the main argument used to prove the hypothesis is determined to be false since there are no requirements for a hypothesis to be true or false to be considered scientific or not scientific.

As a matter of fact, there are many scientific hypotheses and theories currently known to be false that we consider scientific because they meet the requirements imposed by the scientific adjective to explain their empirical observations.

So, let's not confuse the scientific nature of a hypothesis with our ability to generate hypothetical tests that could yield evidence to prove it true or false because being scientific or true or false are different attributes of a hypothesis with very different determination criteria.

Therefore, even though some of you may think that the ID hypothesis is scientific because it can be falsified, that's probably not a good reason to reach this conclusion due to the logical inconsistencies inherent to this falsifiability concept.

And even when we may agree or disagree entirely or to some extent with some or all of the materials presented, this should not affect the determination of its scientific nature for the reasons previously exposed.

Therefore, based on our previous discussion, I think it will be a reasoning error and a waste of time to use this concept to try to "extrapolate" the scientific nature of the ID hypothesis since it will not add any value to the overall analysis and understanding of this hypothesis.

CHAPTER 12

The Paradigm Shift of Existence

Now, I can't walk away from the previous Blue Winged Unicorns falsifiability example, ignoring the fact that most of us struggle with grasping the concept of being able to prove the non-existence (Law, 2011) of the Blue Winged Unicorns outside of the boundaries of the surrounding areas of the large tree in the backyard, or the non-existence of Bertrand Russell's teapot (What is Russell's teapot?, 2022) orbiting the sun between Earth and Mars.

So, I have decided to detour a little bit from the main subject of this work, to bring a different point of view about the existence and non-existent nature of things. To this effect, I propose the following definition of existence as a reference framework to explain why it should not be that difficult to prove a negative or explain the non-existence of something in our observed reality.

New Definition of Existence

1. In our observed reality (space-time dimension), existence (E) is an attribute we assign to physical matter that we can observe

and recognize (not only visually) directly or indirectly through our physical senses.

2. The existence attribute has a logical value of "1 = Existent" for the physical matter we can observe or a logical value of "0 = Non-Existent" for the physical matter we can't observe.

3. The existence attribute of the physical matter has a unit of measure of a physical location (XYZ) at a given moment in time (T); for example, XYZ & T = t.

4. The following three conditions must be met to determine the existence attribute of a given physical matter in our observed reality.

    a. The observer has to be able to directly or indirectly observe and recognize the physical matter.
    b. The observed physical matter must be located within the effective range of the observer's capabilities.
    c. The physical matter being observed and the observer must be present simultaneously.

5. If the above three conditions are met, then the existence attribute of a given physical matter is determined to be:

   E = 1 @ XYZ & T = t

   And from the observer's point of view, there is evidence that the matter in question exists in his observed reality.

6. If any of the above three conditions are not met, then the existence attribute of a given physical matter is determined to be:

   E = 0 @ XYZ & T = t

And from the observer's point of view, there is no evidence that the matter in question exists in his observed reality.

In simpler words, existence is an attribute that we assign to physical matter <u>in our observed reality</u>, and it is determined by the ability of the observer to observe the subject matter and recognize it, in a given location, at a given moment in time.

So the existence or non-existence of "something" should not be such an abstract concept as long as we realize that it is a property of the physical matter in the space-time dimension of our observed reality in the "eyes" of a capable observer.

Then, why does it seem so difficult to prove the non-existence of "something"?

From the logical point of view, proving the non-existence of something should be as easy (or as complicated) as proving the existence of that something. And this is the case because existence (E) is a binary attribute of the physical matter determined by the above definition. So, suppose you can determine the existence of something at a given location and time. In that case, you should also be able to determine its non-existence at the same place and time.

So, why does everyone think that proving the nonexistence of something is not possible?

I think that the primary reason why proving the non-existence of something in our observed reality seems to be so difficult for almost everyone is that most of the time that we are asked to prove the non-existence of "something," the assumptions are that;

- A. we are capable of observing it.
- B. we know what it looks like; therefore, we can recognize it.
- C. we know where (location) and when (time) to look for it.

But unfortunately, we are never provided with enough information to determine if that "something" is non-existent. Therefore, our minds get confused trying to solve an equation with too many unknown variables.

For example;

I have been asked many times to prove the non-existence of many gods. And this challenge always comes after a lengthy argumentation process when a believer "runs out of bullets" trying to prove the existence of his god, and as a last resource, he tosses the burden of the proof over my shoulders to prove him wrong.

But the truth is that we got to this point precisely for the same reasons that I'll not be able to prove the non-existence of his god, or any god!

Because;

    A. He doesn't know if he is capable of observing it.
    B. He doesn't know what it looks like and how to recognize it.
    C. He doesn't know where (location) and when (time) to look for it.

So, my usual reaction to this specific type of challenge usually goes along the following lines:

On the assumption that I know what your god looks like and that I should be able to recognize him through my physical senses, there is no evidence of its existence in the room where I'm writing these notes. Therefore it is non-existent. But maybe he was existent yesterday or is existent today, or will be existent tomorrow in a different room of the house at a different time, but so far, we haven't met.

Now you may argue that because I'm unable to observe and recognize your god at my actual location and time is not evidence of his non-existence since he could exist in "another location" and possibly at a "different time." And I must admit that based on my definition of existence, that is possible in our space-time dimension since he could be in another place now.

But I can also argue that you are asking me to prove its non-existence because you don't know how to prove its existence in the first place. After all, you don't know what it looks like, how to recognize it, or where and when to look for it.

So, we can't determine if he is existent or non-existent in our observed reality not because he is "existent or non-existent" but because of the same basic need for information to be able to reach this determination.

It is evident that our impossibility to determine what we are supposed to look for, whether or not we can observe it, and where and when to look for it creates confusion. But the root cause of the problem is not in answer to this question but in the poorly constructed question that ignores the necessary information to determine the existence or non-existence of something in our observed reality.

It reminds me of the many times my wife yells across the living room while watching TV and asks me to bring her "that" without any more information about what is "that" and where it is located.

A similar challenge occurs, for example, with the speed attribute we assign to physical matter. If you ask someone to determine the speed of an object in our observed reality, you assume that this person can observe and recognize the object. But it would help if you let the person know, as a minimum, where the object is located at a given moment (its existence). This way, he can observe two known locations at different times in its moving trajectory to calculate the distance and the time it took to travel to determine its speed.

The same happens with the acceleration attribute we assign to the physical matter. We need to know at least two locations where the object exists to determine the speed difference over this period.

Without being able to determine the existence of this object in our observed reality at different times and locations, the speed and acceleration attributes of the object are as difficult to determine as their own existence due to the lack of information.

And the same happens with any other attribute that we assign to the physical matter in all the branches of science that depend on the existence attribute of the physical matter.

I know this may sound unfamiliar to some of you who honestly believe that your god exists in the observed reality of our space-time dimension. But if this is the case, then anyone who claims the existence of his god in our observed reality should be able to answer these questions:

A. Are you capable of recognizing him?
B. What does he look like?
C. And where and when can you observe him?

Chances are that you will be as confused as someone who is asked to prove the non-existence of your god. And this is due to the same reasons explained above, lack of adequate information. Because if you don't know if you can recognize him, what he looks like, and where and when to look for him, it will be impossible for you to determine if he is existent or non-existent in our observed reality.

Of course, we can always transfer the nature of the above questions into a different reality or dimension inside the complex nature of our minds. However, this is only an escape from our observed reality that will leave all the above questions unanswered.

Therefore, determining the existence or non-existence of "something" in our observed reality should not be a difficult task to complete, as long as we are capable of observing and recognizing (not only visually) that "something," and we know where and when to look for it.

I'm almost sure that with our current technological capabilities, we should be able to observe and recognize whether or not Bernard Russell's teapot exists between Earth and Mars if he can tell us at least at which distance from the sun his teapot is orbiting.

CHAPTER 13

Testing a Historical Event Hypothesis

How can we gain some confidence that what is being proposed in this historical event hypothesis accurately describes our observed reality? What kind of testing can confirm its proposed explanations are accurate?

Due to the reasoning mode used (abductive) to build the arguments of most of these hypotheses, the only viable option to test them is by comparing their explanations with already known facts of other competing explanations to choose the best possible answer amongst them then.

But the best possible explanation of a series of competing explanations is and will always be an explanation. And no matter how good that explanation can be, the lack of testable predictions will not produce the evidence to prove 100% of the proposed cause.

Therefore, these types of past events' hypotheses generally have lower confidence levels in their conclusions about the responsible cause for a given empirical observation.

However, we can certainly build some confidence in the comparison method employed by ensuring we adhere to a consistent/standardized set of guidelines to help us minimize cognitive biases and have better confidence in its conclusion.

CHAPTER 14

Comparing Competing Hypotheses

There have to be many options to compare competing hypotheses. But certainly, none of them are bulletproof regarding explanations that use the abductive reasoning mode to choose the best cause of a past historical event.

These types of explanations usually lack objective evidence for their claims and have some inherent challenges that must be addressed to produce a valid result from these comparisons.

For example,

> What happens if the compared hypotheses are not competing explanations for the same epistemic end of the empirical observation?

> Or what if we miss or ignore an essential hypothesis in the comparison analysis?

> Or what if comparing these competing hypotheses is not carried out with an adequate methodology?

It is not difficult to see that the effectiveness of comparing competing hypotheses for this type of historical event depends on a few basic but essential requirements that need to be met to gain enough confidence about the certainty of the inference of the best explanation.

> First, the hypotheses being compared have to be competing explanations for the same epistemic end of the empirical observation being explained. Otherwise, it makes no sense to compare them in the first place.

> Second, all currently known competing hypotheses about the epistemic end of the empirical observation need to be part of the analysis. Otherwise, the comparison risks missing a critical explanation, which could be the best.

> Third, a detailed analysis of all the relevant attributes of the competing hypotheses with their corresponding evidence needs to be done using a systematic methodology. Otherwise, the comparison risks reaching biased conclusions due to missing information or evidence.

Suppose the analysis compares hypotheses that do not explain the same epistemic end of the empirical observation, misses an essential competing hypothesis, or does not use a systematic methodology to evaluate all relevant attributes and evidence. In that case, our confidence in the inference of the best possible explanation will not be good enough. And the hypothesis can't be considered an accurate description of our observed reality, being the worst-case scenario of all possible scenarios, an analysis of competing hypotheses that fail to meet these three requirements.

But as a general rule, a well-organized/structured method of comparison that takes into consideration the above requirements helps us present the hypotheses materials in an organized logical order to facilitate its evaluation, and it has the best possibility of meeting everyone's expectations to build consensus and agreement on the best possible explanation.

The Comparison Logical Framework

Before we analyze the comparison method presented by ID advocates in their materials, it is essential to establish an appropriate logical framework within the boundaries of the assumptions contained in the information provided by ID advocates in their scientific proposition.

These assumptions will help us maintain our focus throughout the analysis and minimize potential cognitive biases throughout the comparison process.

Intelligent Design vs. Intelligent Designer
As previously stated, there is a clear difference between an intelligent design and an intelligent designer as it refers to their attributes and capability to act in our observed reality.

Quickly reviewing the topic, an intelligent design is a depository of information, a blueprint of specifications. And as such, it does not possess any attributes or capability to carry out any physical actions in our observed reality without the intervention of a capable entity. Therefore, human beings and nature can't be the result of intelligent design but the result of an intelligent designer.

Intelligent Designer Existence
ID advocates claim that because we exist, and abiogenesis hypotheses have not been able to produce the type of complex information contained in our cell's DNA from the inert matter at random, this proves that the Intelligent Designer was the only

viable cause that existed and was present at the time that human beings and nature were created (Causal Existence).

But this Causal Existence condition can only be met if the analysis is carried out with hypotheses that compete for the same epistemic end of the empirical observation without missing any competing explanations. Otherwise, the intelligent designer's past existence and his act of creation is a hypothetical scenario.

And from the logical point of view, any argument that relies on an abstract concept or a hypothetical scenario or situation (i.e., the Intelligent Designer) as if it was a concrete fact or a real thing commits a **Fallacy of Misplaced Concreteness** and an **Argument from Ignorance** Fallacy, due to the lack of independent evidence to support its conclusion.

Creation Process
When ID advocates say that human beings and nature are the creation of an intelligent designer, they do not provide any specific information about the intelligent designer's process used to accomplish this goal.

But we can assume by elimination that whatever process or method was used by the intelligent designer to create human beings and nature, it had to be different than the method used by members of the same species to procreate new individuals (biogenesis).

And this should be the case because mating and procreating new members from the same species do not require an intelligent designer's presence and intervention.

Human Beings

When ID advocates talk about the creation of human beings, they are referring to us, the most intelligent entity known over the face of planet Earth. And not about other types of entities that could exist outside of planet Earth.

And that the creation of the first human being was done here on planet Earth as a fully functional entity.

The ID Empirical Observation

The empirical observation that ID advocates attempt to explain with their hypothesis is the genetic information in the cells of all carbon-based living entities. More specifically, the programmed instructions that define the physical characteristics and attributes of the type of biological entity that we are.

Now it is important to note that when we talk about the genetic information contained inside our cell's DNA (Deoxyribonucleic Acid (DNA) Fact Sheet, 2020), we are not talking about the actual physical structure of the DNA double helix (chromosomes) made of chemical building blocks called nucleotides that formed by a phosphate group, a sugar group, and one of four types of nitrogen bases (ATGC).

What we are talking about is the <u>logical organization (order and sequence)</u> of the coded patterns of the nitrogen base pairs that define the biological instructions that cells of multicellular organisms like us use to produce the proteins that allow us to execute all the complex functions that support life as we know it.

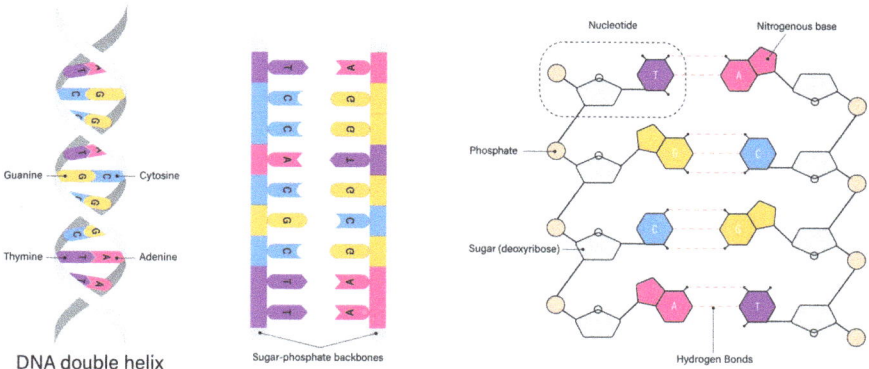

Figure 10 – DNA Deoxyribonucleic Acid

These complex logical patterns or instructions are the information we associate with an intelligent entity's mind (thoughts).

As an illustration, to help us identify the empirical observation explained in the ID hypothesis, we are not talking about the single NPN silicon transistor nor the physical, logical structures that we build from this basic transistor to create all the hardware building blocks of a fully functional 1 MEG USB memory drive.

We are talking about the large amounts of information stored inside the USB hardware media as "1's and 0's", coded as ASCII characters (ASCII Code - The extended ASCII table, 2022) in logical patterns or instructions.

ASCII - Binary Character Table

Letter	ASCII Code	Binary	Letter	ASCII Code	Binary
a	097	01100001	A	065	01000001
b	098	01100010	B	066	01000010
c	099	01100011	C	067	01000011
d	100	01100100	D	068	01000100
e	101	01100101	E	069	01000101
f	102	01100110	F	070	01000110
g	103	01100111	G	071	01000111
h	104	01101000	H	072	01001000
i	105	01101001	I	073	01001001
j	106	01101010	J	074	01001010
k	10?	01101011	K	075	01001011
l	108	01101100	L	076	01001100
m	109	01101101	M	077	01001101
n	110	01101110	N	078	01001110
o	111	01101111	O	079	01001111
p	112	01110000	P	080	01010000
q	113	01110001	Q	081	01010001
r	114	01110010	R	082	01010010
s	115	01110011	S	083	01010011
t	116	01110100	T	084	01010100
u	11?	01110101	U	085	01010101
v	118	01110110	V	086	01010110
w	119	01110111	W	087	01010111
x	120	01111000	X	088	01011000
y	121	01111001	Y	089	01011001
z	122	01111010	Z	090	01011010

Figure 11 – ASCII Binary Table

This is the kind of information that intelligent entities like us use to represent and communicate our thoughts in the forms of text files, audio, pictures, videos, or computer programs with complex logical and mathematical operations.

ID and Abiogenesis, competing hypotheses?

Perhaps the biggest misconception committed by ID advocates in their scientific proposal is the concept of competing hypotheses (Schupback, 2016), which is the root cause of the logical fallacies committed in their argument.

This misconception is very evident in the type of descriptions and explanations that the abiogenesis and the ID hypotheses postulate, attempting to answer different questions in the same causal chain of events necessary to explain the presence of life on planet Earth.

Let's see why.

Whether it happened here on planet Earth or somewhere else in the universe, either by chance or on purpose, there is no doubt that inert matter was transformed into carbon-based living entities at one point in time. Reasons why amongst others, when all living species on planet Earth can no longer sustain life as we know it, they decompose and return to the essential elements of matter from which they were formed.

So, it is difficult to disagree that life, as we know it arose from inert matter since all the empirical evidence we have in this regard tells us that this was the case. What we don't seem to agree with due to our lack of knowledge on the subject is when, where, and how this matter-to-life conversion process happened.

And working primarily on the assumption that this matter-to-life conversion happened here on planet Earth, abiogenesis hypotheses (Rogers, 2023) are attempting to find through several proposed methods or processes **HOW** the existing chemical and environmental conditions on planet Earth approximately 3.5 billion years ago, could have triggered the random formation of simple carbon-based compounds and life forms.

But, it is also evident that these abiogenesis hypotheses are not proposing nor making any attempts at all to prove **WHO** if anyone was the responsible cause for the creation of the genetic information (GI) contained in our cell's DNA.

While on the other hand, ID advocates are attempting to prove **WHO** (intelligent designer) was the responsible cause for the creation of the genetic information (GI) contained in our cell's DNA.

And it is also evident that they are not proposing nor making any attempts to explain the existing chemical and environmental conditions on planet Earth 3.5 billion years ago, nor any mechanism, method, or process that could have triggered **HOW** simple carbon-based compounds and life forms could have emerged from the physical matter at random.

These are fundamental differences between these two types of hypotheses and the questions they are attempting to answer about the origin of life on planet Earth.

HOW those carbon-based compounds and the first living cell emerged randomly from physical matter on planet Earth is a different question than **WHO** was responsible for creating the genetic information in our cell's DNA.

HOW is searching for a random process that could have produced simple carbon-based compounds and life forms from inert matter in the early days of the existence of planet Earth, while **WHO** is searching for an entity or a cause responsible for the creation of the complex information contained in the cell's DNA?

These are, without any doubt, not the same epistemic end of all the possible causal sequences of events necessary to explain the existence of human beings and nature on planet Earth.

Now, without knowing all the remaining causal chain of events in the line of actions required for our existence, there is no coherent/logical explanation for our presence here because all of them are dependent and necessary events to explain our existence.

So, it is not logically sound to compare two different explanations in the same causal chain of events for the existence of an empirical observation to choose one of them as the best explanation while discarding the other as false or unnecessary.

It makes more sense to compare hypotheses pursuing the explanation of the same causal event to choose the best explanation among them.

For example, comparing all abiogenesis hypotheses that are trying to explain how simple carbon-based compounds and life forms could have emerged at random from physical matter to choose the best possible explanation amongst them. Or comparing hypotheses that are trying to explain who or what was the responsible cause of the genetic information contained in our cell's DNA to choose the best possible explanation amongst them.

But how can we choose the best possible explanation for the existence of an empirical observation between two dependent causal events in the same line of actions necessary to explain the existence of the empirical observation?

For sure, explanations of different kinds trying to answer different questions in the same causal chain of events necessary to explain the existence of a given observation do not compete. Because it is impossible to choose one of them as the best explanation for the existence of the observation, discard the other as false or not necessary, and maintain a logical sequence of events and a sound rationale for the existence of such observation.

So, contrary to what ID advocates believe, the intelligent design and abiogenesis hypotheses do not compete for the same epistemic end of the same empirical observation because they attempt to explain two different events in the same causal chain of events necessary to understand our existence.

And this is the root cause of the **Non-Sequitur** logical reasoning error committed by ID advocates in their main argument when trying to compare two non-competing hypotheses for the existence of the genetic information contained in the cell's DNA of carbon-based entities.

Therefore, due to this logical reasoning error, the comparison analysis between the ID and abiogenesis hypotheses does not produce any useful or valid information. And much less advances our quest to answer **HOW** simple carbon-based compounds and life forms emerged at random from physical matter or **WHO** or what was the responsible cause for the creation of the genetic information contained in our cell's DNA.

Based on this, there is no logical justification to go any further into the rest of the ID materials to prove that this hypothesis has not been adequately argued by its proponents and that any conclusions reached throughout the comparison of these non-competing hypotheses are logically flawed.

However, as discussed before, this does not mean that the conclusion of the ID argument is false. It only means that the argument presented by its advocates to try to make their point about the responsible cause for human beings and nature on planet Earth by comparing these two non-competing hypotheses is invalid.

In my opinion, the ID hypothesis is an attractive proposition considering that there are currently no other competing hypotheses attempting to explain the cause or causes responsible for the emergence of the genetic information in our cell's DNA. But it requires further research and a different type of argument to prove it right.

The pyramids of Egypt

A down to Earth example of competing hypotheses can help us better understand the previous subject.

Think about the Giza Pyramids of Egypt (Pyramids of Giza | National Geographics, 2017). When you are standing before them, several questions come to your mind, one behind the other. Why were they built? When were they made? Who built them? How were they built? Etc. All of them are valid questions about the many necessary causal events that ultimately ended in the magnificent view of the existence of the pyramids.

But does it make sense to compare a possible explanation about **HOW** these pyramids were built with a possible explanation about **WHO** built them to choose the best explanation for their existence?

Or to compare a possible explanation about **WHEN** were the pyramids built, with a possible explanation about **WHY** they were built, to choose the best possible explanation for their existence?

It seems evident that the answers to these questions don't address the same need for knowledge (or epistemic end) about the existence of the pyramids because they explain different and dependent causal events necessary for the eventual existence of the pyramids in our observed reality.

Therefore, although the answers to the above questions are all related to the existence of the pyramids, it is evident from the type of descriptions and possible explanations that they are not competing to explain the existence of the pyramids because these are causal events in the same line of actions necessary to explain their eventual existence.

Hence, comparing any two of these non-competing explanations in the same line of actions to choose the best while discarding the other as false or not necessary leaves us with a fragmented/incomplete sequence of events to explain the existence of the pyramids.

Then, it seems more logical to choose competing explanations that answer the same need for knowledge (epistemic end) about the same causal event related to the pyramids and choose the best explanation amongst them.

For example, comparing explanations that address how the pyramids were built to choose the best explanation.

ID and Abiogenesis, the only competing hypotheses?

By comparing these two non-competing hypotheses, ID advocates assumed that the genetic information in our cell's DNA could only be explained by the creation of an intelligent designer or by the spontaneous emergence at random from inert matter through a random abiogenesis process.

Nevertheless, somehow they managed to ignore other competing causes that can explain the empirical observation of this genetic information without the need for a willful act of creation of an intelligent designer or the spontaneous emergence of life from inert matter, incurring in **False Dilemma** fallacy.

Then, we must ask ourselves if it is strictly necessary to know the mechanisms responsible for transforming inert matter into living organisms to determine the cause or causes of genetic information in our cell's DNA and our eventual emergence here.

Is our presence here possibly due to a cause other than a random abiogenesis process or an intentional creation process to convert inert matter into living organisms?

Is there an alternate process capable of producing the same empirical observation of the genetic information in our cell's DNA without converting inert matter into living organisms through unknown processes or methods?

It seems to me that the fact that we are a living entity over the face of the Earth, formed from physical matter found on planet Earth (and many other places in the universe), does not necessarily mean that the only potential explanation for our presence here is that we came into existence from inert matter here in planet Earth.

Is it possible that the first instance of this matter-to-life transformation happened elsewhere other than here on planet Earth?

The lack of adequate evidence for this effect should not prevent us from considering other possible causes that do not require a physical conversion from inert matter to living organisms on planet Earth as potential explanations for the presence of the genetic information in our cell's DNA and our eventual existence here.

The Biological Entity

Many years of empirical observations of the natural world show that only living organisms can give life to other organisms through natural biogenesis reproduction; this is the underlying principle of the universally accepted Cell Theory (Bailey, 2020) that all living organisms are made of cells, the basic unit of life's structure, and cells only arise from preexisting cells. And this is confirmed by the fact that our DNA is irrefutable evidence of the inherited evolutionary history from our biological ancestors (Stanford, Allen, & Anton, 2009).

And because of this gained body of knowledge, we can assert that every living organism that exists on planet Earth was brought to life by another living organism (parents) through the synthesis of chemical compounds and structures through a biological reproduction process.

So, it is no secret that every one of us is the child or offspring of our parents through this natural reproduction process, which is tested every single day, all around the world, every time a mother gives birth to a new child.

Therefore, if we are a living entity composed of cells, there is a very high probability, based on years of empirical observations, that we are the offspring or child of another ancient biological entity whose origins are unknown today. And that as a result of this, a copy of his genetic information was transferred to our ancestors, who transferred it to our parents, who transferred it to us.

Then, based on our knowledge about the biological reproduction process of the species, we can't discard the possibility that the first human beings on planet Earth could have been the result of a natural procreation process that originated somewhere in the universe and was later brought here.

And this possibility can very well explain 100% the presence of the empirical observation of the genetic information we inherited in our cell's DNA and the eventual existence of our species on planet Earth.

In other words, the genetic information in our cell's DNA could result from natural procreation rather than unknown matter-to-life conversion processes like the one used by the proposed Intelligent Designer or the abiogenesis hypotheses.

However, what is interesting to note is that even though there is no conclusive evidence to prove that carbon-based living entities originated here on planet Earth, we seem to be locked into the idea that the only possible explanation for our existence here is an unknown matter-to-life conversion process. And we selectively rule out the evidence and knowledge about this biogenesis procreation process when looking for answers about the origins of the genetic information contained in our cells and our eventual emergence as a species here.

Now, that's what I call a real bias!

But where do we draw the starting point of this ancestral hereditary chain of events that eventually resulted in the emergence of all carbon-based living entities on planet Earth? And at what level of biological complexity, a single cell (LUCA) or a completely functional living entity?

Let's think for a moment. We know that there was no life on planet Earth at some point due to the general chaos and forces of nature since environmental conditions were not favorable for sustaining life as we know it.

But based on the fact that we are here now, it is evident that at a subsequent point in time, the environmental conditions of planet Earth improved and became more friendly and capable of sustaining life. Therefore, it must have been during this second period, when the environment was not harmful to life, that an unknown event triggered life into Earth.

Could life on planet Earth, rather than emerging from inert matter through abiogenesis or a creation process, have originated elsewhere in the universe and was later brought to Earth by other unknown causes? (Was God an Alien? | Unveiled, 2022)

Well, even though the topic of extraterrestrial visitors on planet Earth has been the subject of many past and present conspiracy theories, it would be very naive from our end to pretend to know as a fact without any supporting evidence, that we are the only planet in the universe that can foster and maintain some life. Especially when we already know that there are many exoplanets (Pat, 2021) in many regions of the Milky Way galaxy made up of elements similar to those of the planets in our solar system with favorable environmental conditions to foster and sustain life as we know it.

So, is it possible that an extinct or existent exoplanet of our universe at some point in time harvested some biological entity that evolved and ventured just like our species is starting to do these days to travel outside their home planet to explore the universe that we share?

The truth is that there is no way of knowing that we were the first planet to foster life in our universe. Hence the possibility that other planets in our universe could have done this first can't be discarded, and it has to be assumed. And although we can't prove this possibility to be 100% true due to our technological limitations and the need for more objective evidence, we can't either prove it to be 100% false for the same reason; therefore, it is still a plausible scenario.

And it is possible that these visitors from other places of the universe could have brought with them the basic building blocks of life to colonize our planet or even procreated their species here. And due to this biological procreation process (biogenesis), this long/ancestral hereditary chain of events gave birth to our species here on planet Earth.

We may all argue about the cause of visiting biological entities from other parts of the universe due to the need for adequate evidence. And also because our generalized assumption due to cognitive bias is that we are the only living intelligent entity in this vast universe.

But I guess we can argue the same about the intelligent designer's cause because this cause had to be present here on planet Earth to carry out his act of creation. And with all possibilities came from a place of the universe other than planet Earth, which by definition makes him also an extraterrestrial entity.

So, what's the difference between the proposed intelligent designer (ID) and my proposed biological entity (BE)?

From the logical point of view, there is virtually no difference because both of them are inferred causes for the existence of human beings and nature on planet Earth through our abductive reasoning mode.

And this is the case because, in the same way that we infer the potential past existence of an intelligent designer (ID) through the empirical observation of the complex genetic information contained in our cell's DNA as it relates to our intelligence attribute, we can also infer the potential past existence of a biological entity (BE) through the empirical observation of the complex genetic information contained in our cell's DNA as it relates to our capacity to procreate our species through a biogenesis process.

The main difference is that the past existence of the BE as the responsible cause for the genetic information contained in our cell's DNA can be established by the large amounts of available evidence that proves that only life can give life (Bailey, 2020) and our current knowledge of the reproduction process of all carbon-based living entities through the synthesis of chemical compounds and structures through a biogenesis process (Amit, 2021). In contrast, the past existence of the ID as the responsible cause for the genetic information in our cell's DNA can't be established by choosing the best possible explanation of two non-competing hypotheses through an unknown process of creation.

Therefore, since there is no conclusive evidence today to prove that this matter-to-life conversion process happened here on planet Earth, we can't discard the possibility that it could have happened somewhere in the universe first and was later brought to our planet by visitors from these places. And that the genetic information in our cells results from a hereditary process from another species, not the result of an abiogenesis or a creation process here on planet earth.

But somehow, we keep looking for the origin of life on planet Earth only as the result of unknown matter-to-life conversion processes, ignoring the possibility that we could be the result of a well-known biogenesis procreation process.

Then, we can conclude that the deliberate creation of an intelligent designer is not the only potential cause that could explain the genetic information in the DNA of human beings and nature as ID advocates claim because all available empirical evidence about the biological reproduction of the species (biogenesis) suggests that a proposed biological entity (BE) is also an adequate cause that can explain the presence of this genetic information and our eventual existence too.

And as with any other hypothesis, it should not be easily discarded without adequate research and should be considered a competing explanation for the empirical observation of the existence of the genetic information contained in our cell's DNA and our eventual emergence on planet earth.

The Biological Entity Argument

Now, what kind of argument can be used to prove the past existence of a biological entity as a potential explanation for the presence of genetic information in our cell's DNA?

Well, let's try the following argument:

Premises

Rule
All living biological entities are the offspring of previously living biological entities.

Cause
Human beings are living biological entities.

Conclusion

Effect
Therefore, human beings are the offspring of a previously living biological entity.

As you can see, this proposed BE argument is a deductive reasoning mode argument.

To the best of our knowledge, both premises (rule and cause) are true as of this day. And the conclusion (effect) necessarily follows from these two premises as a valid conclusion.

Then as explained in the Logic 101 chapter, this deductive reasoning argument can only be proved wrong by disproving either of its premises. For example:

1. If anyone can prove that not all living biological entities are the offspring of previously existing biological entities.

Or,

2. If anyone can prove that human beings are not living biological entities but something else.

But suppose none of the above two premises can be proved wrong or false. In that case, the conclusion about human beings being the offspring of a previously existing biological entity must be considered valid.

At first glance, this seems to be a challenging argument to defeat, at least in short to mid-term. But science is all about time and knowledge. And there is a possibility as we learn more about the origin of carbon-based living entities, the role that natural causes like evolution could have played, as well as the opportunity that our actual understanding of what we consider to be a living organism (or entity) can evolve to something different than what it is today, that this argument can eventually be proved to be wrong. In the meantime, I think that it will "hold water" for a while.

Now, if you think identifying a proposed biological entity as the responsible cause for the genetic information in our cell's DNA is a **Blind Authority** or **God of the Gaps** fallacy, think again! Because you are probably not aware that you are a biological entity with this "supernatural" power to procreate your species, which is a well-known natural cause for science, something that we can do and science can explain very well. Hence, there is no supernatural causation in inferring this cause.

Is the BE the ID?

As discussed, the BE cause can explain the empirical observation of the genetic information in our cell's DNA. And its past existence is established through the association of the presence of this complex information in our cell's DNA and the current knowledge that we have of the capability of our species to procreate new individuals through a biogenesis process by replicating parents' DNA materials (Pray, 2008) without the need or intervention of unknown matter-to-life conversion processes like abiogenesis or creation.

However, this does not explain how the proposed BE entity from another place in the universe existed before we did. But it can certainly explain how humans emerged on Earth for the first time.

But let's not confuse ignorance with impossibility because this is fallacious (**Argument from Ignorance**). Neither the fact that we don't know how or when this proposed biological entity emerged in our observed reality is not evidence to prove that it never existed, nor it prevents us from being able to find this in the future. It is only evidence that today we don't know.

And the same is true for many other causes that we don't know how or when they emerged in our observed reality, but we can infer their previous existence by the effects they left behind.

And, because there is no evidence to contradict this proposed biological entity cause as the one responsible for the presence of the genetic information in our cell's DNA, as remote as we might think it can be, we can't discard the possibility that the best explanation for the presence of this information in our cells and our existence here on planet Earth is that we are the product of a natural procreation process (biogenesis) of a previously existing biological entity, that most likely came from another place in the universe other than planet Earth.

Now, ID advocates may argue that if we inherited the genetic information in our cell's DNA from a biological entity outside of planet Earth, then "someone or something" was responsible for the creation of that biological entity and that this proves that we are the result of the creation of an intelligent designer.

But this argument is fallacious in principle because it only transfers the scope of the ID hypothesis from human beings and nature on planet Earth to a different biological entity in an unspecified part of the universe invoking supernatural causation that can't be explained with their current argument, which in effect, brings us back to square one!

And suppose ID advocates are willing to accept the possibility of the existence of an unknown biological entity in other parts of the universe from a logical point of view; in that case, there is no valid reason to reject the possibility that we could very well be descendants of a biological entity from another part of the universe since all available evidence about the reproduction of the species (biogenesis) proves that we are the offspring of previous generations of biological entities, which explains the same, epistemic end of the empirical observation of the presence of the genetic information in our cell's DNA.

However, we can't overlook that we are formed from inert matter that exists on planet Earth and other parts of the universe. And that our existence here is evidence that this matter-to-life conversion process happened for the first time at some point in time, somewhere in the universe, either by chance or on purpose. But we still don't know the cause or causes that triggered this event, nor when or where it happened for the first time.

And much less do we know the remaining and necessary causal chain of events that allowed those first living cells that resulted from this conversion process to become the complex biological entity we are today.

Now, let's go "back to the future" and extrapolate one more step.

Let's say that there is a possibility that this proposed BE was smart enough and knew the secrets of life that we are so desperately looking to create life from inert matter. And rather than procreating his species here (biogenesis), he opted to create a new kind of biological entity or entities from inert matter to help him adapt to Earth's survival challenges.

Hence, it is possible that the process used by this BE to create life as we know it from inert matter could be the same process that abiogenesis advocates are so eagerly trying to find, as well as the same process of creation that ID advocates mentioned in their scientific hypothesis.

Then, under this alternate scenario, it is not difficult to imagine that such a biological entity capable of exploring the universe, colonizing other planets, and even creating life from inert matter could very well be the proposed intelligent designer that ID advocates have inferred through the intelligence attribute of our species. Since to be able to accomplish all of these actions, this biological entity had to be (or maybe still is), among many other things, an extraordinary designer.

Now, I recognize that this proposed scenario may look like an exciting script from someone's imagination to create a science fiction movie. Primarily because of our "unexplained" cognitive bias about the possibility of the existence of any entity from outside planet Earth, which reminds me of the struggle Nicolaus Copernicus[8] went through when he placed the Sun, rather than the Earth, as the center of the solar system.

But as good scientists, we should not discard this possibility without having some evidence to contradict its premises because the empirical evidence of the genetic information in our cell's DNA and our current knowledge in the bioinformatics field of science tells us that the past existence of this proposed BE could be the responsible cause for the genetic information in the cell's DNA and the eventual emergence of our species on planet Earth.

Of course, there is still a possibility that we can eventually find an abiogenesis process that can yield the basic building blocks of life (cell) as we know it in a relatively short to medium period. But this will only explain the first event of a very long sequence of causal events necessary to explain our existence as the fully functional intelligent species we are today. As well as there is a possibility of the existence of another unknown "supernatural" cause that we can't explain today with our current state of knowledge, which could be responsible for the spontaneous creation of our species as a fully functional entity.

But both of these two possibilities, as of today, still lack adequate evidence to support their claims.

---

8 Copernicus: Facts, Model & Heliocentric Theory - HISTORY - HISTORY

And if we are ever going to find out how this proposed biological entity / intelligent designer accomplished these goals either here or somewhere else in the universe, one way or the other, we will have to follow the footprints that were left behind in our historical DNA structure by the first atoms that initiated the incredible journey to transition from cold, inert matter to simple life forms, to the complex biological entity that we are today; an enterprise journey that is going to take many future generations of successive intelligent species like ours to complete.

But most importantly, this has to be done before our species steps on top of those ancient footprints with our increasing knowledge in the fields of bioinformatics (Bioinformatics, 2022), genetic engineering, quantum computing, and artificial intelligence, confounding the paths of our past emergence and natural evolution with our future course of "unnatural" selective evolution.

Genetic Manipulation

Once we have established through a natural biogenesis process the possible existence of an unknown biological entity from another place in our universe as a potential cause for our presence here, there is also a possibility that such an entity, capable of traveling through space, was able to adapt to his survival needs in our environment by manipulating the DNA of other biological entities (plants, animals...etc.) that were existing here or brought from elsewhere during these visits, opening the door to an almost infinite number of possibilities for new species.

And although we can't be 100% sure about what processes or methods could have been used by the proposed biological entity to manipulate the DNA of those species, at least we have as a reference today's knowledge of genetic engineering (Smith, 2022) technologies that we (a biological entity) have developed to manipulate the DNA materials of the species in planet Earth.

These genetic engineering processes, such as recombinant DNA (rDNA), CRISPR (Mateuszi & Sharma, 2020), etc., don't happen in nature alone. But it is something that we engineer in a laboratory test tube to create DNA sequences that would not otherwise be found in an organism's genome. Then we propagate this modified DNA to other organisms like bacterial cells, yeast cells, plants, animals, and even humans to enhance or alter the characteristics of their genetic makeup.

These genetic engineering processes (8 Tools and Techniques of Gene Manipulation, 2018) have numerous applications in medicine, research, industry, agriculture, and environmental management that help us perpetuate the survival of our species.

For example;

> In medicine, for insulin production, human growth hormones, cure for genetic disorders, vaccines, and many other drugs.
>
> In research, to modify the genetic composition of organisms to be used to discover the functions of specific genes.
>
> In the Industry, for the mass production of proteins.
>
> Crops or organisms with higher nutritional value and herbicide resistance are created in agriculture.
>
> And in the environment to produce a specific type of bacteria capable of producing biodegradable plastics.

At least, we know, based on our knowledge about genetic engineering, that it is possible to manipulate the DNA of the species to favor the expression of desired physiological traits or the generation of selected biological products (Nutrition, 2022) to help us adapt to our survival needs on planet Earth.

And having these genetic manipulation technologies at our disposal, I will venture to say that it is very likely that we will use them to modify the genetic composition of crops, animals, and even the humans that will be sent to Mars (mars.nasa.gov) on the next 50 years to guarantee the success of the first human colony in this planet.

Therefore, if we can accomplish this today, it is not a far-fetched possibility that a biological entity visiting planet Earth a long time ago could have done the same to guarantee its survival on our planet.

Then, it is possible that our DNA information was the result of a genetic manipulation process of other biological entities from planet Earth or other places in the universe.

In other words, in addition to being the product of a natural biological procreation process (biogenesis) within members of the same species, we could also be the product of genetic manipulation in the hands of a capable entity that combined properties of different species, including but not limited to the proposed biological entity itself into a new kind of specie.

But how can we know if a genetic manipulation process was used in the past to conceive one of the distant ancestors of our evolutionary chain? Is it possible that there is still some traceable evidence of this event somewhere on planet Earth?

If this is the case, then this type of evidence will strengthen the possibility of the proposed biological entity as the responsible cause for the presence of genetic information in our cell's DNA and the eventual evolution of human beings and nature through a genetic manipulation process.

Well, as a possibility, a plausible one. But chances are that if there is any evidence of this past event somewhere in this world, the best place to start looking for it is inside our historical DNA structure.

The Comparison Method

Generally speaking, we don't like the odd sensation we experience when we see something new that we can't explain. It is like getting a non-maskable interrupt request to our central processing unit telling us to override the priority of all other tasks currently under execution in our random-access memory to execute a high-priority task to find a logical explanation for what we see.

And without even looking for adequate evidence, we instinctively use our cumulative knowledge base and the abductive reasoning mode to generate possible explanations about the cause or causes that could have produced the observation.

Then we make a quick "like/don't like" mental comparison of one or two of the most critical aspects of the explanations and choose the one that "we feel" is the most appropriate before leaving the scene to continue with our everyday life; this is probably an acceptable approach in situations where we only want to satisfy our curiosity and need for knowledge on subjects that don't have significant implications in our day-to-day life endeavors.

But what if the observed subject is of critical interest to national security? And the chosen best explanation will determine whether or not we should deploy some troops in a faraway desert in the Middle East to protect our nation's interests abroad.

It seems reasonable to think that the more that is at stake or risk about the given observation, the more thorough and complete this analysis needs to be so that we can gain a higher confidence level about the credibility of the chosen explanation and the adequacy of the corresponding actions to be taken.

So, it is fair to ask ourselves how thorough/complete the analysis of the competing ID hypotheses needs to be to gain some confidence about the best explanation. And what is at risk if we do a poor or incomplete analysis and choose a wrong answer?

In this case, it is not about deploying troops or declaring war on a foreign nation. But it has everything to do with whether or not the proposed ID hypothesis, as presented and argued, is the best explanation for the presence of the genetic information contained in our cell's DNA and the eventual existence of human beings and nature on planet Earth.

As discussed before, we already know that ID advocates compare their hypothesis with known abiogenesis hypotheses under the wrong assumption that these are competing hypotheses for the same epistemic end of the same empirical observation. So, comparing these non-competing hypotheses, no matter what method is used or how good it is, will yield illogical conclusions due to the previously mentioned fallacies and reasoning errors. And this is why I will not spend time or effort reviewing and debating every aspect of the competing hypotheses analysis presented by ID advocates.

However, from the didactic point of view, it is worth reviewing The Explanatory Power of a Hypothesis method used by ID advocates in their scientific proposal to gauge the effectiveness of this tool in the following areas:

    A. Structuring and organizing the information to be compared.

    B. Identifying the most important attributes that need to be compared.

C. Identifying the available evidence that supports each attribute being compared.

D. More importantly, determining the decision-making algorithm to measure the explanatory power of a hypothesis to choose the best possible explanation.

These are undoubtedly essential aspects that need to be taken into consideration when analyzing competing explanations of past events if you want to minimize the natural cognitive bias that we all have toward favoring our explanations vs. someone else's.

Now the need to make a meaningful analysis of the above aspects makes it almost necessary to employ a structured method to collect, organize and analyze all of the information from the competing hypotheses.

Otherwise, sorting through all this information without a well-defined methodology becomes a formidable intellectual challenge hindering our ability to maintain an organized and unbiased logical thought process.

Fortunately, there are several structured methods or guidelines (with known strengths and weaknesses) that can be used to perform structured analysis of competing hypotheses (Uuganbold, 2007) that involve a relatively high degree of uncertainty like the ones that result from our abductive reasoning mode to identify causes of past historical events.

Among which the most commonly used by many analysts in various fields of government and science (data mining, cognitive psychology, visualization, probabilities, and statistics) is the Analysis of Competing Hypotheses (ACH) developed by Richards J. Heuer, Jr. (Andres, 2022), of the CIA in the 1970s. This analysis was created to help the

CIA agency minimize cognitive biases and make decisions on subjects with a high risk of reasoning errors.

Adherence to these methods provides a broader span of control over the available knowledge base of information on the competing hypotheses and a more impartial evaluation by testing every proposed argument and available evidence against each hypothesis to determine the most supported explanation.

Nevertheless, it is not a trivial task; it requires discipline and a considerable amount of time and resources to use.

The Explanatory Power of a Hypothesis

Now, ID advocates opted to do the comparison of their hypothesis and the abiogenesis hypotheses using the concept of the Explanatory Power of a Hypothesis (Explanatory power, 2022) within the context of the Philosophy of Science (Understanding Science, 2022), which is the ability of a hypothesis as presented to explain the subject matter to which it pertains effectively.

But for most hypotheses, the ability to explain their subject matter is primarily determined by whether or not the documented test results in its body of evidence match the expected hypothesis predictions. Something that we have to recognize, as discussed before, is a difficult task to achieve when working with hypotheses of past events that are trying to explain causes for which there is virtually no practical experimentation that can be carried out to validate its predictions.

However, scientists and philosophers of science propose using the Explanatory Power of a Hypothesis within this context as if it was a measurable characteristic of a hypothesis that can be effectively used for comparison purposes. Like the Explanatory Power of a Regression (Valchanov, 2021), a statistical analysis of a hypothesis value representing how much of the total variability contained in a given set of measurements can be explained by the existing relationship between dependent and independent variables of the regression.

However, outside the statistical analysis field of collecting, exploring, and presenting large amounts of data to discover underlying patterns and trends, the concept of the explanatory power of a hypothesis within the context of the philosophy of science still has a few challenges that need to be addressed to yield some practical results in comparing scientific hypotheses of past events.

Let's review this concept in a little bit more detail.

As a general rule, it is commonly accepted amongst scientists and philosophers that a hypothesis with great explanatory power makes few assumptions, has significant predictive power, and helps reduce the uncertainty of the subject matter precisely and accurately.

While a hypothesis with poor explanatory power makes too many assumptions, has lower predictive power, and does not help to reduce the uncertainty of the subject matter in a precise and accurate way.

At the conceptual level, it seems reasonable to think that the above-proposed relationship between these variables and a given hypothesis can give us a general indication of how convincing a given explanation is.

However, this general level of abstraction is not good enough to make an effective comparison of competing hypotheses in the "real world," and it is usually necessary to break them down into more discrete/measurable elements that could be used for comparison purposes, in such a way that it is generally accepted that a given hypothesis has more explanatory power than another if;

    A. It has fewer assumptions.

B. It depends less on authorities.
C. It depends more on observations.
D. It has more facts or observations.
E. It changes more "surprising facts" into "a matter of course" (following Peirce 2).
F. It provides more details about what should be expected to be seen.
G. It provides more details about what should not be expected to be seen.
H. It provides more details of causal relations of the description.
I. It is more falsifiable, according to Popper.

A quick examination of the above elements shows that all of them are discrete variables[9] with values somewhere between zero and a positive value for the number of instances of the occurrence of each component in the proposed hypotheses.

Two decrease the explanatory power as the number of instances increases (A & B). And seven of them increase the explanatory power as the number of instances increases (C to I). At the same time, two of them (C & D) seem to be confounded by the same variable (observations), making it necessary to differentiate between the number of observations and those known to be true or facts.

A. The number of assumptions. (Less is better)
B. The number of dependences on authorities. (Less is better)
C. The number of dependencies on observations. (More is better)
D. The number of facts or observations. (More is better)
E. The number of "surprising facts" changed into a "matter of course." (More is better)
F. The number of details expected to be seen. (More is better)
G. The number of details not expected to be seen. (More is better)

---

9 What are Discrete & Categorical Variables? | Types & Examples of Discrete, Categorical & Continuous Variables - Video & Lesson Transcript | Study.com

H. The number of details of causal relationships. (More is better)
I. The number of falsifiability tests. (More is better)

But how do we go from the discrete value of the number of instances of each variable to the magnitude of the total explanatory power of a given hypothesis?

Should the total explanatory power be a straightforward addition of all variables' contributions?

Or should each one of these variables have a different relative contribution or weight for the total explanatory power? And if so, what guidelines can determine this relative contribution?

It seems to me that those variables that can be verified with objective evidence, like the number of assumptions, dependence on authorities, facts, or observations, should carry a more significant contribution or relative weight than those variables that are more subjective, like the number of details expected to be seen, details not expected to be seen, number of surprising facts changed into a matter of course or the number of falsifiability tests.

In addition, there is also a need for an algorithm to determine the hypothesis's total explanatory power, considering the positive or negative effect of each one of these elements and their respective contribution or relative weights.

There is also a possibility that the above basic assumptions on calculating the explanatory power of a given hypothesis can differ for different types of empirical observations, which imposes the need for some guidelines to help us adapt this method to different needs.

Now, considering that it has been challenging to get answers to these fundamental questions and find good information, references, or examples on how to use this explanatory power method, it seems reasonable to conclude that although it makes sense at a conceptual level, the use of this concept for comparison purposes at a practical level lacks the necessary definition to become a practical comparison tool.

And while this lack of definition or structure allows for a great deal of flexibility to new hypotheses' proponents, it is also a double-edged sword that opens the door to a higher degree of cognitive bias.

Because this lack of structure and definition allows this explanatory power to become a relative measurement that can be easily confused, for example, with the amount of information presented for a given explanation; this can foster or promote lengthy expositions and discussions of information in favor of the hypothesis supported by the proponent making the comparison, and less description and discussion of the information of the other competing hypotheses.

Nevertheless, suppose a given format and algorithm to measure the magnitude of the total explanatory power of hypotheses is agreed between its proponents; in that case, it should be possible to carry out a reasonable comparison whose confidence level on the final determination of the best explanation will be adequate for hypotheses with significant differences on explanatory powers. But inadequate for hypotheses with similar magnitudes. This is due to the inherent resolution in some of the elements this comparison tool measures.

Therefore, based on the fact that the analysis of competing hypotheses presented by ID advocates is comparing two non-competing hypotheses and that it missed other important competing explanations for the same empirical observation of the genetic information, as well as the use of an inadequate comparison method to reach its conclusion, we can conclude that our confidence level on the intelligent designer cause as the best explanation for the presence of the DNA information in our cells, as well as our presence here on planet Earth is very low as argued by its advocates.

CHAPTER 15

Is the ID hypothesis a theory?

Generally speaking, a hypothesis is an assumption, a guess, a conjecture, or an idea that proposes a tentative explanation of an observed phenomenon in the natural world, so that it can be tested to determine if it is an accurate description of our observed reality.

While a theory is a well-substantiated explanation commonly regarded as correct based on a body of evidence that has been repeatedly confirmed through experimentation and observations and used as principles to explain and predict the observed events of the reality in which we all live.

A theory is the most reliable, rigorous, and comprehensive form of knowledge and the closest thing we have to what we all recognize as truth. It differs from a hypothesis because a theory is not a tentative explanation of a phenomenon in the natural world but an accurate, reliable description to the best of our knowledge of our observations of our reality.

And while it is a commonly accepted practice (although not correct) in our day-to-day social conversations to use the word theory within the context of an argument implying that something is a non-proved supposition, it is not acceptable to use this interpretation within the framework of scientific work. Since this is considered to be a contradiction that allows the use of two different/opposing meanings for the word theory, one meaning a conjecture or guesswork, and the other representing a proven fact or a truth.

Incurring this type of contradiction is considered an **Ambiguity/Equivocation** fallacy that creates confusion within the framework of scientific work. It makes it very difficult to draw a line between what is considered a truth and what is recognized as a lie. A practice that is very common through the ID materials with the many references from its advocates to their hypothesis as if it was a proven scientific theory.

So, when discussing a scientific subject, we must be careful with the word theory to ensure we use it within the proper context of our arguments. Otherwise, the credibility of the scientific work and its conclusions is undermined.

Then, based on the previous analysis, we can conclude that the ID hypothesis, as currently presented and argued by its advocates, is still a scientific hypothesis (not a theory) that requires further development to "iron out" the many fallacies and reasoning errors that undermines its ability to be considered as an accurate description of our observed reality.

However, I think that the possibility of the intervention of an intelligent entity somewhere in the chain of causal events necessary for the existence of human beings and nature on planet Earth is plausible, and it should not be discarded based on an inadequate argument.

Maybe a different point of view and a different argument about the presence of this genetic information in our cell's DNA could bring adequate evidence to get one step closer to the cause responsible for the origin of the cell of carbon-based living entities.

CHAPTER 16

The Paradigm Shift of Information

Once more, I'm faced with the need to challenge another traditional concept that has been around in the scientific community for a while without an apparent consensus about what it is. But suppose I want to move forward to complete this work; I have no choice but to bring the information concept to the table so that it can be adequately analyzed and defined within an adequate reference framework to use it effectively in the remaining part of this work.

So the following pages are dedicated to exploring what I think is information and how this relates to our ability to think and communicate to others in our observed reality.

What is a thought?

As a necessary condition to introduce this new concept of information, we need to review what a thought is.

Thoughts (Simran, 2021) are an elusive concept widely used in many science disciplines with similar but different meanings. They relate to many human skills like planning, organizing, interpreting, and making decisions. They are intimately connected with our ideas, opinions, beliefs, and emotions that are part of our consciousness, reasons why it is probably a challenge to create a universal definition of thought that can satisfy the many disciplines of science in which this concept is applied.

But without detouring into the labyrinth of the many definitions and interpretations of what can be considered a thought, I propose that we focus primarily on the dimensions of this concept as it relates to us human beings.

1. Thoughts are recognized/individual acts of human beings.
2. They are the product of our ability or skill to think.
3. An external environmental stimulus can trigger thoughts through our physical senses.
4. An internal stimulus from our cumulative knowledge in response to our thinking process can trigger them.
5. Thoughts are a free flow of ideas that allow us to imagine, meditate, reflect, and reason.
6. Thoughts have physical and logical boundaries. Physical boundaries are determined by our abilities to sense and observe, and logical boundaries are determined by intellectual capabilities (intelligence, knowledge) to interpret observed behaviors.
7. We, as individuals, have certain degrees of freedom about what we can choose to think of.
8. We can also relate or associate different ideas with what we think of.
9. Based on this interrelated thinking process about the different things (reasoning) we can choose to think of, we can reach conscious reality-oriented conclusions.

10. However, these conclusions are useless unless we act upon them to derive an observed effect on our environment.
11. If we decide to act upon these conclusions, these empower us to take action over our perceived world or reality.

A quick analysis of the above dimensions of this concept shows us that all thoughts share a common property; they are intellectual responses within the boundaries of our minds, an important characteristic that I will use as a starting point to build the new proposed definition of information.

However, just for reference purposes, I would like to propose one more definition for the concept of thoughts:

A thought is an intellectual response generated within our mind's boundaries to a perceived stimulus from the environment around us or an inferred stimulus from the reasoning process of our collective knowledge.

Notice that the mechanisms by which humans think or generate thoughts respond to the interaction with our environment or the perceived reality. And these, more than often, are overlooked and taken for granted without realizing the sequence of events that triggers this activity within our brains and without recognizing that this thinking process in itself is the foundation by which we freely interact with the environment that surrounds us to make conscious decisions to guarantee the survival of our species.

What is Information?

Information has been a very complex concept to explain and define for many years, and today it is still fertile ground for philosophers and scientists trying to explain it. And this is the case mainly because of the close relationship between information and our thoughts, how they relate to human behavior, and the multiple contexts of life in which we use information.

As a result, many of the available definitions of information have been crafted primarily from the point of view of how the information is used and communicated rather than the point of view of where it originated and how this one is represented in the different available mediums. The result of this approach is the many available definitions of a concept that, if adequately studied, there should be no reason to have more than one definition.

There are probably several approaches to introducing this new definition of information. I can either tell you straight what I think it is and works on the details afterward. Or I can build a step-by-step explanation for it and finish with the new definition as a conclusion, which we can later test against some of the traditional definitions to see how it gages. So, I will choose the long path since it provides a better understanding of this concept.

Can you guess?

Let's start this process with the following mental exercise or experiment. Let's assume that you and I are sitting on chairs facing each other in an empty room with our hands on the top of a table, and I ask you the following question;

    Me: Can you guess what I'm thinking right now?

You will probably be surprised by the question knowing very well that you don't have any psychic ability to read people's minds.

    You: How do I know what you are thinking right now?

Me: I don't know. You tell me. How can I help you understand what I'm thinking right now?

You: Well…, I guess you can tell me, and I will then know. Or if you don't want to tell me, at least you can give me some hints that I can use to try to guess.

Me: O.K., I'm not going to tell you. That's too easy. But I will give you a hint.

Then, I reach into my shirt pocket and pick up a small piece of paper and a pencil and make the following drawing, and hand it to you so that you can see it:

You: Hum. What are you thinking? About love?

Me: That's right! You got it.

Well, that was the end of the experiment. But many things we take for granted in our day-to-day interactions have happened here quickly. So let's review them step by step.

Step 1 – Before I asked you to guess what I was thinking, I was very conscious and aware of the exercise I wanted to do to illustrate my definition of information. And you were looking into my face waiting for me to say something since you had no idea what this was all about.

Step 2 – I realized that if I wanted to carry on with the experiment I had in my mind, I would have to choose a way to communicate with you so that you would become aware and understand the instructions I needed you to follow.

Step 3 – I assumed humans could communicate through several methods using our physical capabilities and bodily senses. For example, we could emit sound waves through the air with our vocal cords and listen to them with our ears, or we can write our Q&A's with characters on a piece of paper that we can both see and recognize.

Step 4 – I also assumed that chances were very high that you were educated in English and that I could use it as our communication protocol to exchange questions and answers.

Step 5 – Then, with the use of my lungs, I created an airstream that provided the necessary energy to stimulate my vocal cords as I modulated them with the muscles in my larynx and amplified the sound that resonated in my throat, nose, and mouth, starting the process of converting the first question that I had in my mind into coded sound waves in the English language that traveled at 343 meter/second in the air (medium) between you and me.

Step 6 – Then, using your ears, you listened to the sound pressure level fluctuations in the air and converted them into electrochemical impulses that traveled in your auditory nerve into your brain. This allowed you to retrieve, decode and understand the first question I transmitted a few seconds before.

Step 7 – We went back and forth between steps 5 and 6 until I decided not to tell you what I was thinking about (no more voice communication) because that was too easy for you.

Step 8 – Then, I stopped to think for a few seconds and decided to use a visual symbol (not English) as a hint to help you guess the thought that was in my mind.

Step 9 – I assumed you would probably be familiar with the universally accepted heart symbol, and I engraved it on the top of the paper's surface with the energy from my hand's muscles and the carbon material from the pencil.

Step 10 – You looked at the piece of paper and recognized the symbol on its surface, which made you think (or guess) that I was probably thinking about love.

Step 11 – You communicated your guess by speaking to me again through sound waves in the English language protocol.

Step 12 – I listened to your answer and confirmed that you correctly "guessed" the thought I had on my mind.

Now, I must admit that is a lengthy explanation for something that happened so fast and that we did so naturally without thinking aloud. But splitting our interaction into each of the above steps allows me to "double click" in some of these steps to make several important observations about the nature of what I think is information.

For example, let's review **Steps 5** and **9**. These two steps are particularly relevant because the thoughts I had in my mind made an effective transition from realms of my mind (or consciousness) into a different type of medium or matter. And this transition was possible by using some of my body's energy to alter the physical properties of those mediums or matter with a known communication protocol for both parties that allowed me to send you my thoughts.

Without this transition of thoughts from my mind to the physical medium or matter and the effective use of a known communication protocol for both of us, you would probably never had a chance to find out what the exercise was all about or even "guess" what I was thinking of.

In **Step 5**, the thoughts that I had in my mind about the instructions that I wanted you to follow were physically transferred by my vocal system to sound waves in the air space medium between you and me, representing my thoughts in the form of sound pressure level fluctuations on this medium.

I guess that if you want to confirm this as a fact and want to "see" those sound waves representing my thoughts traveling through the air, you could connect a microphone to an oscilloscope, talk to it, and look at the corresponding waves on the screen and even determine its amplitude and frequency in the time domain.

Also, as I voiced my thoughts, they were encoded in English, which we both know to communicate. And if I had used a different language, you would probably notice that I was attempting to communicate through voice but would not be able to "decode" and understand what I was trying to say.

While in **Step 9**, the thought that I had in my mind (Love) was physically transferred to a piece of paper with the use of the energy from my hand and a pencil by "engraving" with carbon a heart symbol (different communication protocol) on its surface. And this image was reflected by light at a speed of 299,792,458 meters/second from the modified surface of the paper through the air into the cornea of your eyes. It then reached the optic nerve of your eyes that transmitted the visual representation of my thought in the form of electrochemical signals to your brain. And this is how you figured out what I was thinking of.

Notice that in **Steps 5 & 9,** I had to make use of some energy (airstream from my lungs, the force from my hands) to modify the communication medium (air, paper) with the physical representation of my thoughts so that these could be transmitted and received between both parties.

Notice, too, that the physical modifications performed to the air medium with my vocal system had a shorter duration than those performed with the pencil to the surface of the small piece of paper; however, both were adequate to communicate my thoughts.

Alternatively, we could have our entire conversation (exchange of thoughts) on this experiment by sending text messages of every question and answer back and forth to our cell phones with the use of the energy from our hands and the electromagnetic signals carrying our thoughts through the air (medium) in another known language for both. And we would still be able to accomplish our goal, which proves that a single thought can be physically represented in the same medium (air) with the use of different energy sources(mechanical, sound, light, electromagnetism, etc.) and different communication protocols(languages).

Physical Representation of Thoughts

So, let's review some essential concepts that can be derived from the previous experiment using as a guideline the following diagram:

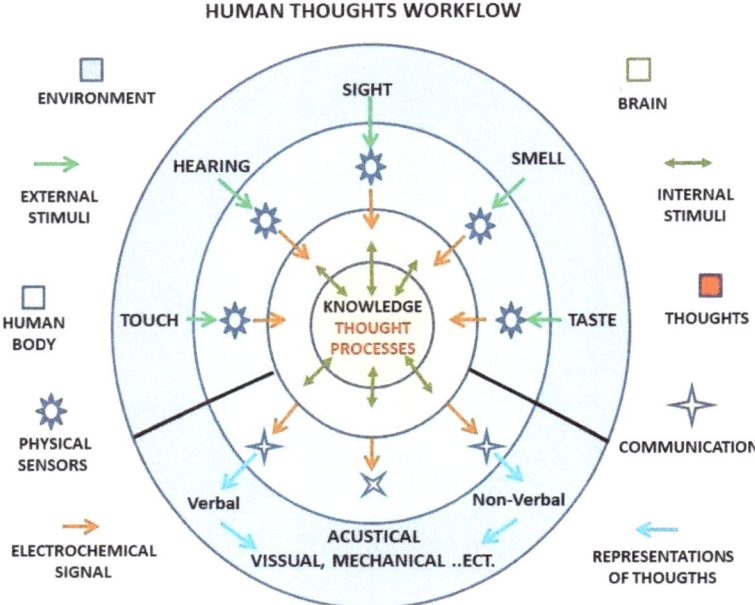

Figure 12 – Human Thoughts Workflow

Our body ☐ exists in a physical environment ☐. We constantly receive external stimuli → from this environment through our physical senses ✻ (touch, hearing, sight, smell & taste) that convert the stimuli received into electrochemical signals → that our brain ☐ is capable of processing and understanding.

Our brain is the depository ○ of our collective knowledge and thought processes. It generates internal stimuli ⟷ that are converted to electrochemical signals → to trigger our verbal & non-verbal communication capabilities ✦. This produces energy and creates acoustic, visual, and mechanical representations of our thoughts →

by modifying the physical properties of matter or medium in the environment.

These modifications that we perform to the physical properties of the matter and medium are used to communicate our thoughts to others, which is a competitive advantage for the survival of the species that allow us to pass knowledge from one generation to another.

Now, expanding on this definition, it is imperative to note that without thoughts, there is no information because this is its physical representation embodied in the matter or medium of our observed reality.

This representation of thoughts is accomplished by applying energy to the physical structure of the matter or medium to create a new form or structure that becomes the container where our thoughts are stored for communication and later retrieval.

Now, our thoughts can be physically embodied in many types of matter or medium as long as these can retain them for enough time to support the intended communication method. And these embodiments can be changed or re-arranged in many different forms, structures, or coding patterns to protect their integrity. However, to communicate them effectively, the sending and receiving ends of this information need to know and understand these forms, structures, and coding patterns to retrieve these thoughts effectively from the media or matter.

Therefore, what we usually recognize as information in our day-to-day endeavors are physical representations of thoughts (ideas, messages, feelings, etc.) embodied into the physical matter or medium using energy. And these physical representations are usually created to retain or communicate our thoughts to others through verbal (voice) and non-verbal (documents, pictures, objects, etc.) means.

However, those thoughts are neither the matter, the medium, nor the energy used to alter the physical characteristics where these are embodied. But they need these elements to be embodied and communicated.

Therefore, based on the previous observations about how thoughts flow from our inner minds to the observed reality in which all live, we can conclude that;

Information is the physical representation of thoughts created on any existing matter or medium of our observed reality, modifying its physical properties or characteristics through energy.

Of course, we all create information for many reasons. But the ultimate and primary reason to create any information in any matter or media is to store our thoughts and knowledge for later retrieval, communication, and action purposes.

Information tests

Now, let's review some of the available definitions of information out there to refresh our minds about what is being said about this concept and compare the previously proposed definition with them.

Information:

DEFINITIONS THAT I DISAGREE

These are some of the definitions of information that I disagree with for the reasons exposed in each one.

1. Defines the nature of an entity and its characteristics.

This is one of the things we can do by using information. We can define entities' nature and characteristics, but this does not explain what information is.

2. Is data or knowledge.

    Data are facts or statistics used for reference and analysis. And knowledge is the awareness and understanding we gain through analyzing and reasoning those facts and statistics.

    You can use the information to retain data or statistics and gain knowledge through its analysis, but data and knowledge are different concepts than information.

3. Help us to understand abstract concepts.

    This is one of the things that information allows us to do; it helps us understand abstract concepts. But it is just another use for information, not an accurate definition of what information is.

4. It Is expressed as the content of a message.

    This tells us one way in which information can be embodied (as a content of a message) for communication purposes, but it does not define what information is.

5. Can be encoded in various forms for transmission and interpretation.

    This tells us that information can be encoded in various forms for transmission (communications) and interpretation, but it does not define what information is.

6. It Can also be encrypted for safe storage and communication.

    This tells us another thing we can do with the information: encrypt it for safe storage and communication, but once more, it does not define what information is. It

7. Is the communication or reception of knowledge or intelligence.

    Knowledge can be transmitted and received (communicated) as information, but the information is neither knowledge nor intelligence. These are three different concepts.

8. Is knowledge obtained from investigation, study, or instruction.

    Knowledge from investigation, study, or instruction can be embodied as information and communicated to others.

    But information and knowledge are two different concepts.

9. Are facts and data.

    Facts and data can be embodied as information for communication purposes, but they are not information.

10. It is the communication or reception of knowledge or intelligence.

    You can transmit and receive (communicate) knowledge or intelligence using information, but the information is not the communication or reception of knowledge or intelligence.

11. Information is the reduction of uncertainty[10]. (Markouski, 2022).

---

[10] Shannon's Theory

Indeed, the more informative a statement is, the more uncertainty it reduces about a particular event. But this is one of the things that can be accomplished with information (reduce uncertainty), but it is not a good definition of what information is.

12. A sequence of characters or arrangements of something that produce a specific effect.

    Information can be represented as a sequence of characters or arrangements of something to enable its communication. Still, as a stand-alone entity, it can't produce any effect in our observed reality. It requires a capable entity to understand and act upon the communicated message.

DEFINITIONS THAT I AGREE WITH, TO SOME EXTENT

On the other hand, I do agree to some extent, but not entirely, with some of the available definitions of information listed below;

13. Is intelligence, news.

    Information is not Intelligence; this is an attribute or characteristic of humans. On the other hand, news can be information because they are the physical representation of thoughts embodied in the physical matter or media (paper, digital media, etc.).

14. The attribute inherent in and communicated by one of two or more alternative sequences or arrangements of something (such as nucleotides in DNA or binary digits in a computer program) that produce specific effects[11].

    I agree that information could be an arrangement of things in a given sequence or order which can be used to communicate an attribute or a message. Still, I can't entirely agree

---

11 https://www.merriam-webster.com/dictionary/information

that information alone produces any specific effect because the information is a depository or an embodiment of thoughts in the matter or media. No effect can be derived from those thoughts without the intervention of an entity capable of understanding and acting upon them.

For example, no effect can be derived from the nucleotides in the cell's DNA without the intervention of the structure of a fully functional cell, or no effect can be derived from any computer program without the intervention of computer architecture capable of understanding and acting upon these program instructions.

15. A signal or character (as in a communication system or computer) that represents data.

    A signal or a character can represent information or data, but information by itself is not a signal or a character.

16. Is something (such as a message, experimental data, or a picture) that justifies a change in a construct (such as a plan or theory) that represents physical or mental experience or another construct.

    This definition of information is probably the one that I like the most, as it relates to us when it refers to a message or a picture that represents a physical or mental experience. It attempts to define what information is from its source and gives examples of how it can be represented in our observed reality.

Therefore, as we can see from the previously widely used/recognized definitions of information, this concept is generally misinterpreted with different descriptions of what can be accomplished with information and how it is manipulated for storage and communication purposes.

But they don't provide a clear definition of what is the information concept.

Hence for all practical purposes, when I make use of the term information throughout the rest of the book, I'm referring, as described before, to the following proposed definition;

> **Information is the physical representation of thoughts created on any existing matter or medium of our observed reality, modifying its physical properties or characteristics through the use of energy.**

CHAPTER 17

An Alternate Argument for the Origin of the DNA Information

I will be of no real help to science in the long term if I make this effort to evaluate the ID materials but don't take the time to make an attempt based on the learning experience that I had to provide an improved argument about the intelligent design hypothesis and help science advance one more step in this direction.

Therefore, I have decided to present a different approach to argue this hypothesis and highlight the challenges I see moving forward to prove this hypothesis false.

Then, if I were to argue the intelligent design hypothesis, I would probably use the following argument:

PREMISES

Rule
Only intelligent entities are capable of creating large amounts of information.

Cause
Human beings and nature have large amounts of information in their cell's DNA.

CONCLUSION

Effect
Therefore, the information contained in the cell's DNA of human beings and nature is the creation of an intelligent entity.

As you can see, this argument is a deductive reasoning mode argument. To the best of our knowledge, both premises (rule and cause) are true as of this day. And the conclusion (effect) necessarily follows from these two premises as a valid conclusion.

Notice that there is no need to make any mentions, references, or comparisons to the existing/known abiogenesis hypotheses about the beginning of life on Earth to reach this conclusion. Because the scope of those hypotheses is to find a process on how inert matter at random under the "right" environmental conditions could have made the transition to carbon-based compounds and simple life forms and not to explain what cause or causes were responsible for the information contained in our cell's DNA.

Also, notice that the conclusion of this argument is not about the creation of human beings and nature on planet Earth as in the original inferred ID argument but only about the creation of the genetic information contained in our cell's DNA. And this is the case because this proposed argument has no evidence in its premises to prove a cause responsible for the creation of human beings and nature but only for the DNA information in their cells.

Now, having a sound argument about an intelligent entity as the responsible cause for the creation of the genetic information in our cells might not be adequate evidence to prove that human beings and nature were also the product of the act of creation of an intelligent entity.

And this is the case because there is a clear difference between what information is and the media where the information is contained. And unless proven otherwise, there is a possibility that the cause responsible for the emergence of the media (cell) where this genetic information is stored is a different cause than the one responsible for the information.

In addition, notice that I'm using the term "Intelligent Entity" instead of "Intelligent Designer" to differentiate the type of argument used for each of these causes. Also, the term "Intelligent Entity" communicates the notion that human beings are also intelligent entities and not unexplained supernatural causes.

And although we don't have all the necessary knowledge to create this type of complex information from scratch, we have to realize that the line that divides our understanding of what an intelligent entity is today or can be tomorrow is a moving target where our current knowledge on this subject ends, and our ignorance begins.

But if someone thinks that invoking an intelligent entity as a responsible cause for the existence of the information in our cell's DNA is a variant of invoking supernatural causation or a **God of the Gaps** fallacy, think again! Because you are probably not aware that you are an intelligent entity capable of generating large amounts of information, and this does not necessarily mean that you are an unexplained supernatural cause to science.

Then, as reviewed in *the Logic 101, Reasoning Modes* section, this deductive reasoning mode argument can only be proved wrong by disproving either premise. For example:

AN ALTERNATE ARGUMENT FOR THE ORIGIN OF THE DNA INFORMATION

1. If anyone can prove that large amounts of information can be generated by another cause or combination of causes other than an intelligent entity, then the argument's conclusion could be false.

Or,

2. If anyone can prove that what we classify today as large amounts of information in our cell's DNA is not information but something else, then the argument's conclusion could be false.

But, if none of the above two premises can be proved wrong or false, the conclusion about the cell's DNA information being created by an intelligent entity logically follows from its premises and must be considered valid.

Keep in mind that although this argument is valid, it does not explain or provide any direct evidence on how this proposed intelligent entity created and stored this genetic information in the cells of all carbon-based living entities. And because of this, we can't completely rule out other potential causes that could have derived the same effect of the presence of this genetic information in the nucleus of these cells.

However, this is probably a good step in the right direction based on the evidence at hand. This could help us start the process of demystifying the origins of this genetic information on all carbon-based living entities because it identifies one potential cause based on the empirical observations of the presence of this information in the cells.

Remember that the previous argument does not prove that an intelligent entity created human beings and nature. Still, it demonstrates that the information in their cell DNA was the creation of an intelligent entity, an important milestone for the eventual emergence of carbon-based living entities.

At first glance, this seems to be a tough argument to defeat, at least in short to mid-term. But science is all about time and knowledge. And there is a possibility as we learn more about the origin of the cell of carbon-based living entities and the original level of complexity of its genetic information, as well as the possibility that our understanding of what we consider to be information can evolve into something different than what I proposed in the previous chapter, that this argument can eventually be proved wrong. But for now, I don't see any good reasons why it should not be considered an accurate description of our observed reality.

Now, there is an undeniable fact about all carbon-based living entities: we are here and alive. And our existence can't be possible without first creating the basic building block of life (cells) where this information is contained.

Therefore, somehow at one point in time, somewhere in the universe (including Earth), either by chance or on purpose, carbon-based cells arose from inert matter. And the process responsible for creating the first carbon-based cell, with all probability, is the same unknown process that ID advocates inferred in their scientific hypothesis and the one that abiogenesis advocates eagerly try to find.

The Parietal Art

The Parietal Art Cave Paintings (Hirst, 2019), with illustrations (information) of extinct animals, humans, and geometric figures made of charcoal, ochre, and other natural pigments on the walls of rock shelters and caves throughout Europe, is the creation of a past biological entity that lived on Earth more than 30,000 years ago.

And although scholars still debate about the purpose of this "cave art" in those days, most scientists agree that an intelligent entity in

Europe created these images a long time, the Neanderthal (History.com Editors, 2019).

However, the fact that we can infer from the evidence left behind that these images were the creation of this intelligent entity does not prove that the material of the walls of these caves and shelters was also the creation of this intelligent entity. Since the existence of these cave materials had a completely different cause a long time before the presence of the intelligent entity that painted these images.

The same happens when we see those old settlements built with mud and rocks by ancient civilizations. We can infer by their organization and structures that an intelligent entity made them. However, this does not prove that the materials used in their construction were created by the same biological entities that built them.

Similarly, being able to explain the cause responsible for the existence of the genetic information in our cells does not necessarily explain the cause responsible for the emergence of the cells where this information is contained.

CHAPTER 18

Genetic Information vs. the Cell

Then, due to the clear difference that exists between information and the media where the information is contained, my proposed alternate argument about an intelligent entity responsible for the creation of the information in our cells might not be a good argument to prove that it was an intelligent entity the one responsible for the origin of the cell of all carbon-based living entities.

Working from this assumption creates a paradox similar to the chicken and egg paradox; what came first, the information in the cell's DNA or the cell? And it also incurs a **Questionable Cause Fallacy** when it suggests that the origin of the carbon-based cell is the information contained in its DNA without providing adequate evidence to support this claim, just because the information in the cell's DNA is regularly associated with the cells and plays an essential role in its biogenesis reproduction process.

So we have to be careful not to confuse the correlation (or relationship) between the information in the cell's DNA and the cell with their respective potentially different causations for their existence. Because although building a cell indeed requires the information that is stored in the cell's DNA, this statement has been confirmed to be valid only during the biogenesis process, where the cell uses this information to synthesize the chemical compounds and structures of new cells.

But this assumption is irrelevant outside the biogenesis process because it does not explain the origin of the first cell necessary to initiate this reproduction process. Therefore, searching for the cause of creating the first cell that triggered this biogenesis process is still necessary.

And this is why science has made a clear distinction between creating the first cell from inert matter (abiogenesis) and reproducing a cell from another cell (biogenesis) to avoid the contradiction created by the paradox of the information contained in the cell's DNA and the cell.

So let's not assume that the cause of the origin of the cell of all carbon-based living entities is the genetic information contained in the cell's DNA without having objective evidence.

CHAPTER 19

Can information alone create a new cell?

To bring closure to the paradox of the DNA information and the cell, we have to either prove or disprove the possibility that the information in the cells could generate the first carbon-based cell.

To do so, we have to keep things in the proper perspective.

The information contained in the cell's DNA, as discussed before, is the physical representation of the thoughts of an intelligent entity that was created by modifying the logical organization (order and sequence) of the coded patterns of the nitrogen base pairs of the DNA structure to define the biological instructions necessary to build the complex entity that we are.

But once these thoughts are embedded into the DNA structure in the form of double helix chromosomes as a set of coded patterns or instructions (ATGC), they need all the support structures (organelles) of the carbon-based cell to maintain a controlled environment, with enzymes, catalysts, energy...etc., to support the replication and protein synthesis processes required to create a new functional cell (Vedantu, 2022).

Let's glance through some of the basic information to illustrate what we mean by this.

All carbon-based living entities are categorized into two main groups based on the difference in their cell structure, the Eukaryotes or Prokaryotes (Roger, 2022), (Academy, 2022).

Eukaryotes can be single-cell or multiple-cell organisms and include animals, plants, fungi, and protists, while Prokaryotes are single-cell organisms that include bacteria and archaea.

Eukaryotes and Prokaryotes organisms have cell membranes to protect the cell's content by controlling the movement of substances into and out of the cells. Inside this cell membrane is the Cytoplasm, whose primary function is to support and suspend organelles and cellular molecules inside the cell. Both types of cells contain ribosomes which play a vital role in assembling proteins.

Eukaryotes cells have a nucleus containing DNA (genetic information). In contrast, Prokaryotes cells lack internal membranes and no nucleus, and their DNA floats freely inside the cytoplasm.

Eukaryotes cells, with some exceptions, also have other structures called organelles, including mitochondria (energy exchangers), Golgi apparatus (secretory device), endoplasmic reticulum (membrane canals), and lysosomes (digestive apparatus);. In contrast, prokaryotes cells lack all these organelles (see Figure 13).

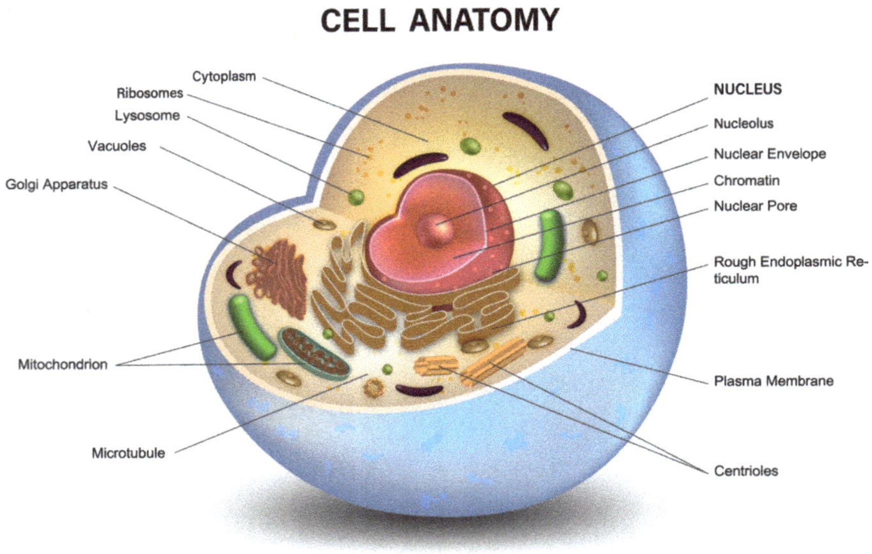

CELL ANATOMY

Figure 13 – General Eukaryote Cell Structure

These carbon-based organelles of the Eukaryotic cell are necessary to support and maintain a controlled environment to allow the production of two daughter cells from a parent cell through a Mitosis (Encyclopaedia Britannica, 2023) process. And without this basic Eukaryote cell structure and organelles, the genetic information in the Eukaryotic cell nucleus can't perform the cell division process to create new cells.

Therefore, the alternate argument that I presented about the cause responsible for the creation of genetic information of the cell, although it is a valid argument, does not provide any supporting evidence for the emergence of the first cell of carbon-based living entities as ID advocates subliminally implied it in some of their materials. Because you can have DNA with some genetic information on it, but new cells will not be reproduced without the Eukaryote cell structure and organelles.

Hello World!

In a parallel analogy to help illustrate the above topic, you can be the author (or creator) in a small Notepad text file (**Hello**) of the following basic JAVA program.

```
HelloWorld.java - Notepad
File  Edit  Format  View  Help
/*
 * First Java program to say Hello
 */
public class Hello {    // Save as "Hello.java" under "d:\myProject"
    public static void main(String[] args) {
        System.out.println("Hello, world!");
    }
}
```

Figure 14 – Hello World Java Program

This program is a physical representation of your thoughts created to display "**Hello World!**" on a computer screen. It was stored (saved) on a USB media by modifying its physical properties with energy so that it can be retrieved, communicated, and executed by others on a computer.

Then it is not difficult to see that the JAVA program stored in the USB drive only contains the physical representation of your thoughts in the instructions necessary to display the "**Hello World!**" message on a computer.

But this stand-alone USB drive in your hands is not capable of performing its intended function unless it is granted access to the logical architecture and functionality of your computer, which proves, as I said before, that information alone is useless unless a capable entity acts upon it to read, interpret, and execute those instructions.

In addition, it is also evident that you can't claim to be the creator of the computer where you plug your USB drive to load the program and run it just because you are the author of this simple JAVA program.

In other words, information alone, as in this example, can't explain the origin of the transistor, which is the basic building block of your computer's functional architecture. And you can't claim to be the creator of the computer hardware architecture just because you wrote the Hello.java program that runs on it.

Similarly, the information in our cell's DNA as a stand-alone unit cannot generate new cells unless contained inside the basic functional structure of a Eukaryotic carbon-based cell. And neither can the creator of this information (intelligent entity) claim that he is the creator of the cell without providing the necessary support evidence to prove this additional claim.

Therefore, unless it can be proved otherwise, there is a possibility that the cause of the origin of the first cell of all carbon-based living entities could be a different cause than the one responsible for the origin of the genetic information contained in its DNA.

CHAPTER 20

Genetic Information vs. Human Beings

Now, when ID advocates conclude that the only possible cause or explanation for the emergence of the genetic information contained in our cell's DNA is the willful act of creation of an intelligent designer, they are also subliminally implying with the same statement that the only possible explanation for the existence of human beings and nature on planet Earth, is the same willful act of creation of the intelligent designer.

However, we already know that they can't justify any of these claims to be true with the inappropriate comparison of two non-competing hypotheses and an abductive argument that contains logical fallacies and fails to establish the past existence of the intelligent designer in our observed reality.

So, we need to ask what kind of evidence exists today that can prove that the emergence of the genetic information in our cells and our presence here on planet Earth was a single event that happened at the same time and was triggered by the same cause as subliminally implied by ID advocates.

Or is there a possibility that these two were separate events driven by different causes at different times and locations? One for the emergence of the genetic information in our cell's DNA and another for our presence here on planet Earth.

As you start to realize, it seems quite possible that the cause responsible for the emergence of the genetic information in our cells does not necessarily have to be the same cause responsible for our presence here on planet Earth. And neither one of them had to necessarily happen at the same time and location as subliminally suggested by ID advocates through the act of creation of an intelligent entity.

And this is the case because the fact that we are a living entity on planet Earth with DNA information in our cells does not rule out the possibility that the emergence of this information could have happened elsewhere outside planet Earth driven by one cause and that our presence here on planet Earth could be the responsibility of a different cause.

Thus, assuming that the cause responsible for the emergence of our cell's information is the same cause responsible for our presence on planet Earth could be a potential reasoning error due to the lack of evidence to support this assertion since it is possible that the causes for these two events in the same line of actions necessary for our existence here are entirely different, rather than the same one.

After all, up to this day, nobody has come up with some real objective evidence to prove that a single act of creation was responsible for the existence of fully functional human beings and other living entities on planet Earth.

CHAPTER 21

Is the cell processing information?

We have a carbon-based cell structure that we can't explain its origins very well with DNA information inside its nucleus that points in the direction of an intelligent entity as a potential cause of its existence.

Then, the most basic question that comes to my mind under this scenario is for what the Eukaryotic cell of all carbon-based living entities needs this DNA information?

But then, I realized that this question had been answered a long time ago by many people since we know very well, based on the many years of research on this topic, the vital role that this genetic information has in the repair of damaged cells, as well as in the growth and procreation processes of new individuals. So, I don't think we will find anything new on this subject by reviewing the carbon-based cell processes that use this genetic information (Laham, 2021).

However, the fact that there is an essential role for this information inside the carbon-based cell from the logical point of view brings a different perspective on the nature of these cells and a potential new avenue to find an answer to its origins (D'Onofrio & An, 2010).

As discussed before, the primary and ultimate reason to create any information in any matter or media is to store our thoughts for later retrieval, communication, and action purposes, which tells us that before any information can serve any purpose, it has to be processed by a capable entity to act upon it; otherwise, it is entirely useless.

And it is evident in this case, based on our knowledge about the biogenesis process, that the carbon-based cell is the capable entity that makes use of and acts upon this genetic information. This brings a fundamental question, is the carbon-based cell an information processing unit?

I think that we all know the answer to this question.

If this is the case, that tells us an essential characteristic about these carbon-based cells that should not be overlooked.

But how can we prove that the Eukaryotic carbon-based cell is processing the information contained in the strands of DNA of its nucleus (Lotha, 2023) without going into too many details about the internal processes that make use of this information? After all, this work is not about a biology class but about the search for the origins of life.

It seems to me that a relatively easy way to prove this assertion is by understanding the modifications that we make to the DNA information of the cells to observe the effect that this has on the characteristics of the resulting individuals. Something that science has done many times when manipulating the DNA of some species to favor the expression of desired physiological traits.

For example, a person's eye color results from the pigmentation of the iris that surrounds the pupil, which is determined by variations in a person's genes involved in the production, transport, and storage of the amount of a pigment called melanin in the front layers of the iris, and this is directly related to the color of the eyes (NIH, 2022).

It is known that two genes (OCA2 and HERC2) located very close to each other in a region of chromosome 15 play a significant role in the production of the P protein that determines the amount and quality of the melanin that is present on the iris.

Then, by controlling (turning it on or off) a region of the HERC2 gene (intron 86), the activity (or expression) of the OCA2 gene is also controlled to reduce or increase the amount of the P protein, which in turn increases or decreases the amount melanin and hence the eye color. A decrease in P protein causes less melanin in the iris and lighter-colored eyes, and vice-versa.

And we know this thanks to the cumulative knowledge we have gained in collecting, classifying, and analyzing biochemical and biological information of the cells for the human genome (The Human Genome Project, 2014).

Then, it logically follows from this sequence of events that the cell has to have some information processing capabilities within the walls of its plasma membrane to retrieve, communicate and act upon these genetic information changes.

This simple test proves that the cell can read, interpret, and act upon the changes we make to its DNA information, demonstrating that it has process information capabilities to follow our thoughts or instructions in the form of physical modifications to its DNA structure.

But this is not a secret. This scenario has been proven to work in the many genetic modifications that we perform to the DNA of the species to favor the expression of some of their physiological traits, for example, GMO foods.

Now, I'll give you a bonus.

The carbon-based cell is not only an information processing unit but also a very sophisticated 3D printing unit, capable of producing new carbon-based cells, organs, and individuals based on the instructions in its DNA. In the same way that today we can make on a 3D printer a three-dimensional part from a plastic filament extruded through a heated nozzle by following a set of programmed instructions created from a digital model.

Therefore, we can conclude that the carbon-based cell is a sophisticated **functional information processing unit** (Schumaker, 2021) and a **3D printing unit** (All3DP, 2023) that creates new biological structures based on the information contained in their DNA. And this is a significant realization that can help us in our quest for the origin of carbon-based cells.

CHAPTER 22

An Argument for the Origin of the Carbon-Based Cell

The implication of recognizing that the cells of the carbon-based living entities are functional information processing units (FIPU) is undoubtedly an essential fact that science should not overlook because this opens the door to infer new potential causes for its origins.

This fact is of particular interest to me because I have spent more than 35 years of my professional career in the industrial manufacturing sector working on the development of hardware and software solutions with all sorts of information processing units that mimic or perform our decision-making process to control the quality, cost, and delivery of products.

And I have always been surprised by the level of complexity and functionality that we achieved through these units to help us capture, store, process, communicate, and act upon the increasing amount of information we have been generating during the past 50 years.

From the first Motorola 6800 micro-controller that I programmed at the university in machine language to control the cycle of a washing machine to the complex UNIX-based OS computer system that I used to control a distributed architecture of computers and PLCs (Programming Logic Controller) programmed in ladder logic used to control the manufacturing process of a high mix, high volume built-to-order process for beepers, all of them were silicone-based electronic devices had functional information processing units capable of reading and executing preprogrammed instructions to achieve the desired manufacturing objectives.

So once I understood the functional architecture and the programing language of these information processing units, I could transfer my thoughts to these units by writing a set of instructions in a specific order and sequence (programs) and loading them into its hardware architecture to achieve the desired results. Then, I left those units running as if it was me (or my thoughts), the one that was directing the decision-making of this manufacturing process.

This was not a trivial task for anyone to do since you had to understand every piece of the unit's functional architecture and every single instruction's syntax and program logic to avoid undesired results. And although I sometimes had to iterate through this process a couple of times, once I had the program thoroughly debugged, the information processing unit followed my thoughts through this set of instructions and executed them right every time.

So, there is no need to go any deeper into this subject to demonstrate that the silicon-based computers that we have created, whose basic building block is the transistor, are indeed very sophisticated functional information processing units whose primary reason for their creation is to help us manage information, take decisions and act upon the things of our observed reality.

But, you can imagine that if programming these information processing units required so much knowledge about these architectures and programming languages, how much more knowledge had the designers of these units to be able to create them?

However, what is interesting is that we are the responsible cause for creating these functional information processing units through our intelligence attributes and the knowledge we have obtained in the different science disciplines. And even more interesting is that there does not seem to exist any other cause over the planet other than us that can create these types or any other type of functional information processing units.

And there is a good reason for this since there is no other living entity over the face of planet Earth that is capable of creating large amounts of information on the physical matter or medium of our observed reality (physical representations of their thoughts), neither capable of creating information processing units to manage and communicate this information with other members of its species.

Think about it! I'll give you a couple of minutes.

In which other places in our observed reality can you identify such functionality and capability to manage or process large amounts of information that has a different cause for its existence other than us, an intelligent entity?

Is it possible, then, that the cause for the creation of the cell of all carbon-based living entities could have been an intelligent entity, in the same way, that we (an intelligent entity) are the responsible cause of the creation of today's silicone computers that help us manage our information needs?

I will venture to say that we can build a sound argument to this effect;

Premises

Rule
Only intelligent entities are capable of creating functional information processing units.

Cause
The cell of all carbon-based living entities is a functional information processing unit.

Conclusion

Effect
Therefore, the cell of all carbon-based living entities is the creation of an intelligent entity.

As you can see, this argument is also a deductive reasoning mode argument. To the best of our knowledge, both premises (rule and cause) are true as of this day. And the conclusion (effect) necessarily follows from these two premises as a valid conclusion.

Then, as reviewed in the *Logic 101, Reasoning Modes* section, this deductive reasoning mode argument can only be proved wrong by disproving either of its premises. For example:

1. The argument's conclusion could be false if anyone can prove that there is another cause or causes other than an intelligent entity capable of creating functional information processing units.

    Or,

2. If anyone can prove that the cell of all carbon-based living entities is not a functional information processing unit but something else, then the argument's conclusion could be false.

But, if none of the above two premises can be proved wrong or false, then the conclusion about the cell of all carbon-based living entities being the creation of an intelligent entity logically follows from its premises, and it has to be considered a valid conclusion.

Remember that although this argument is valid, it does not explain or provide any direct evidence of how this intelligent entity created the cell of the carbon-based living entities. And because of this, we can't completely rule out other potential causes that could derive the same effect of the carbon-based cell from the existing matter in our observed reality.

However, this is probably a good step in the right direction based on our current state of knowledge and the evidence at hand to start the process of demystifying the origins of the carbon-based cell because it does identify one potential cause based on the empirical observations of the cell's structure and functionality.

CHAPTER 23

From Matter to Life

As I approach what I think is the end of this work, one more topic keeps running in circles in my "Logical Processing Unit," and I'm not sure that I completely understand if there is a good explanation for it. But, I will certainly be very sorry if I don't take the opportunity to discuss it, especially when I think it is much related to the current paradigm and comprehension of the origin of the carbon-based cell.

During the course of this effort, I have learned that all abiogenesis hypotheses, among which the three most plausible ones are the lipid world, the protein world, and the RNA world, are attempting to find a process by which the elements of non-living matter (Carbon, Nitrogen, Hydrogen, Oxygen, Phosphorus, and Sulfur) under the "right" environmental conditions, at random could have made a transition to simple carbon-based molecules and life forms.

But often, we look for answers to our most challenging questions with an inadequate reference framework that makes us spin our wheels in circles at the wrong place and moment.

I say this because we know that the biogenesis reproduction process of carbon-based living entities produces new carbon-based living entities (Life from Life). But we never ask ourselves where the matter and the energy required to form these new entities' structures come from.

However, it seems obvious, based on the principles for the conservation of matter and energy (National Geographic, 2023), that when a carbon-based living entity procreates offspring, the raw material and the energy needed to form the body of this new entity comes from the elements of the matter that the mother has supplied to his child through the gestation period, reasons why the bodies of these newborn entities are also formed from the same elements of matter that their parents (What is the human body made of?, 2020).

In other words, the new offspring's body is formed from the elements of matter and the energy in the form of nutrients supplied by his mother, which were converted to new carbon-based cell structures through this biogenesis reproduction process.

Therefore, based on these observations, we can conclude that one sure way to convert the non-living elements of matter into carbon-based living entities, which is the main goal pursued by all abiogenesis hypotheses, is to have an information processing unit with the proper structure and functionality like the carbon-based cell, and an appropriate set of instructions or genetic information to guide and control this matter-to-life conversion process.

Then I wonder; what are the reasons that abiogenesis advocates have to search for this matter-to-life conversion process under those special random environmental conditions when we already know, based on empirical observations, that carbon-based cells are capable of converting these elements of matter to new carbon-based cell structures through this information-guided and controlled biogenesis process?

Why are we looking for a random process to build or create carbon-based cells when we already know a process that has been proven to successfully convert inert matter to life under the guidance and control of the carbon-based cell structure and its genetic information?

Do abiogenesis advocates know something we don't when they are sticking to this random environmental condition for the matter-to-life conversion process they are trying to find? Especially when it is evident, based on our empirical observations of the biogenesis process, that this complex matter-to-life conversion process needs something more than the random occurrence of the unknown effects of many unknown variables at many unknown places and moments in time.

I'm not sure that I completely understand the reasons for this. Still, it seems that we are taking a significant detour from the next logical step if we want to find a process capable of converting the inert elements of matter into carbon-based living cells.

I say this because if we know that the carbon-based cell is capable of converting the elements of matter into new cells and living entities through a biogenesis process, then we should be trying to understand the physical and chemical composition of the building blocks of the carbon-based cell's structure at the atomic level. And not using a "try an error" approach that drives us into the probabilities (Siegmund, 2023) field of science where it is virtually impossible to predict all the environmental conditions and biochemical variables that could influence the outcome of our test predictions.

And this approach becomes particularly intriguing when the assumption is that it took nature billions of years to achieve this and that the so-called "random" requirement of these efforts does not seem so random since all of them are guided by us, an intelligent entity.

I think our focus should be to learn how to build a carbon-based functional information processing unit architecture (or cell) with its corresponding set of instructions (genetic information) to guide and control this matter-to-life conversion process.

Now, based on what we know about the conservation of mass and energy principles, and the biogenesis procreation process, the currently used definition of biogenesis (Life from Life) might not be an adequate description for this observed phenomenon, especially when life means different things on different disciplines of science. So, I will propose that we replace this traditional definition of biogenesis with the following one;

Biogenesis is an information-guided and controlled process that occurs inside the cells of carbon-based living entities to convert the elements of matter into new carbon-based cells and structures.

This new definition provides a better description of what biogenesis does and brings into perspective the importance of genetic information (guide and control) in creating new carbon-based living cells.

Well, maybe it is about time that we leave behind some of our traditional (biased?) preconceptions about the beginnings of life on planet earth (abiogenesis) and once more join our collective knowledge and resources to lead a multidisciplinary team of experts to demystify and understand at the atomic level, from a different perspective, the physical composition, structure, and functionality of all the remaining functional blocks (organelles) of the carbon-based cell. Just in the same way that we joined efforts in 1990 to create the human genome project to identify, map, and determine the sequence of all the genes of the human genome from the physical and functional points of view.

In the long run, this effort will probably help future generations of human beings figure out how to build (or create) and program this and other types of biological information processing units to develop applications that guarantee the survival of our species.

CHAPTER 24

Back to the Origin of Life

Going back to our initial objective in the first chapter about the origin of life, we have to conclude, based on the previous analysis, that the ID hypothesis, <u>as initially presented and argued</u> by its advocates, does not prove the existence of an intelligent designer that acted upon the things of our observed reality at a given location and moment in time to derive the effect of the cell of the carbon-based living entities.

However, <u>as presented and argued in Chapters 17 and 22,</u> this hypothesis gives us two valid arguments to infer the existence and capability of an intelligent entity cause responsible for creating the carbon-based information processing unit (cell) and its DNA information.

And although an intelligent designer, by definition, is also an intelligent entity, I differentiate between both of these causes by using different names to reference the different types of arguments used by each.

Notice, however, that neither one of these arguments provides supporting evidence to identify an intelligent entity as the responsible cause for a direct act of creation of fully functional human beings and nature, but only for the creation of the carbon-based cell and the genetic information.

And this is the case because we have no way of knowing for sure the level of complexity of the first carbon-based cell and its genetic information. Therefore, there is no way of knowing how much of the complex biological entity we are today is due to the creation of the first functional carbon-based cell in the hands of an intelligent entity and how much of it is due to other forces of nature, like evolution.

And only comparing the genetic information (GI) from those early carbon-based cells with our current genetic information could probably answer these questions.

But the earlier fossil records (Rafferty, Fossil Record, 2020) that we have of these first carbon-based cells (cyanobacteria) date back almost 3 billion years (Cooper, 2000), (Somma, 2021) and with all probability, no DNA information can be recovered from those fossils to carry out this genetic comparison.

However, we do have evidence on the fossil record that tells us that we had many different carbon-based living entities in existence in the past, as well as they are in existence today. And they are different because their genetic information is different due to many years of adaptation to a changing environment. And plenty of evidence proves that the Homo sapiens species that we know today is the result of a gradual process of biological changes of earlier species through many years of evolution that contributed to shaping the biological entity that we are today.

In information technology terms, the first carbon-based cells should contain Version 1 of the ".GI" file in their nucleus, and our current (Homo sapiens) cells should have Version 9 of the ".GI" file in their nucleus (see Figure 15).

Therefore, it is probably o.k. to assume, based on this fossil record evidence, that there has to be a significant difference in the genetic information of the first Eukaryotic carbon-based cells and the genetic information contained in our cell's DNA that can't be easily explained with the premises of the arguments in Chapter 17 and 22.

So the fact that the carbon-based cell and its genetic information could have been the creation of an intelligent entity is not enough evidence to prove that human beings and nature were created as the fully functional entities that we have today since evolution could have had a significant impact on the development of those early cells to convert them in the complex biological entities that are in existence today.

But one thing is evident in the chronological sequence of events of life on Earth, and that is that the atomic origin of life predates the carbon-based origin of life for more than 10 billion years, and our origin as a species is relatively young (300,000 years) compared to the first known carbon-based cell contained in the fossil record almost 3 billion years ago (Scoville, 2019).

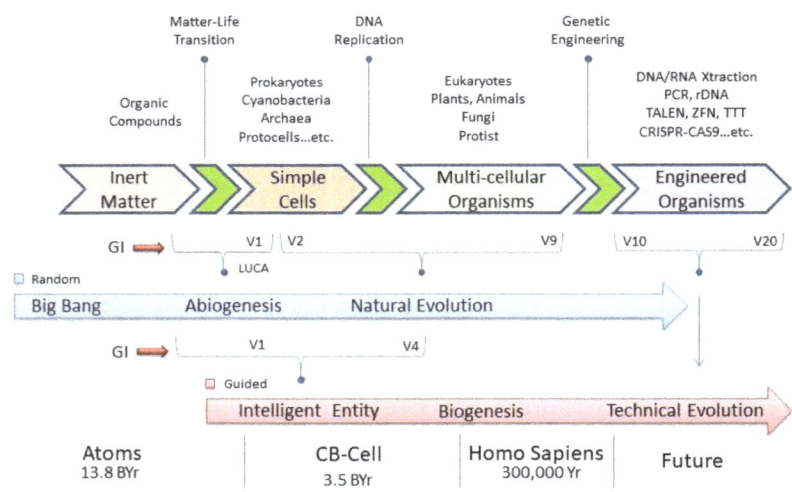

Figure 15 – Life on Earth Roadmap

And because of the large body of evidence contained in the fossil record is very difficult to rule out or argue against evolution's role as a major cause responsible for the development of those early cells into the complex multicellular organisms that we have today.

So, claiming that fully functional human beings and other living entities were the direct results of an act of creation at a given location and moment in time is probably a reasoning error or fallacy known as **Disregarding Known Science**, which occurs when a person makes a claim knowingly or unknowingly disregarding or suppressing well-known science and evidence that weighs against the claim.

But one thing is sure, and that is that the arguments presented in Chapters 17 and 22 make it more difficult to rule out the possibility of the intervention of an intelligent entity as a possible explanation for the emergence of carbon-based cells and their DNA information in our observed reality.

To this effect, I have updated the Origin of Life chart of Figure 5 to reflect this new information set.

The Origin of Life

Definitions of Life	Atomic	Carbon Based
Cause	Big Bang	Intelligent Entity
Existence	Y	Y
Capability	Y	Y
Location	Everywhere	Earth
Time	13.8 BY	2-3 BY

Evidence = Y 　　No Evidence = N 　　Undefined = ?

Figure 16 - The Origin of Life, Atomic and Carbon-Based

This chart shows, as discussed before, that depending on the definition of life that we choose, Atomic or Carbon-Based, there could be two possible origins of life. One was driven by the Big Bang everywhere in the universe some 13.8 billion years ago, and another was driven by an intelligent entity on Earth, possibly 2-3 Billion years ago.

However, remember that this does not mean that one definition of life or its proposed cause is better than the other or that one is true and the other false. Both are probably useful definitions depending on the problem we must solve.

And because there is only indirect evidence of the effects that these causes left behind on other things that we can observe today, we can't completely rule out the possibility of the existence of other causes that could have derived the same effects of the atom or the carbon-based cell from the existing things of our observed reality. So we can't be 100% sure that these two causes were responsible for the effects of the atom and the carbon-based cells in our observed reality.

In other words, as we learn more about the origins of our universe and all living entities, we can eventually find evidence about other causes responsible for the existence of the atoms and the carbon-based cell. But as of today, these are probably good explanations that we can back up with our current state of knowledge.

So, maybe it is about time that we reconsider our current paradigm about the beginnings of life on Earth and we accept the possibility that the creation in the hands of an intelligent entity from somewhere in the universe is also a viable explanation for the existence of the carbon-based cells on planet Earth. And that the evolution of those first cells on the hands of mother nature is also a feasible explanation for the existence of the great variety of complex carbon-based living entities that we have today, and that both of these causes are sequential causes in the same line of actions necessary to explain the origins of all living entities and its actual level of complexity.

Epilogue

Well, it has been a long journey since I started to write this book, and as expected, the world around us has continued to change, sometimes at speeds not expected by anyone under very uncertain times. But in the end, regardless of all the personal challenges I have faced during this time, I was able to meet my commitment to deliver these thoughts to our future generations, feeling that every minute I have spent on this endeavor has been worth it.

I have learned a lot, especially about myself and the comprehension that I developed about the many different topics and disciplines of science in the reality in which we all live.

And while doing so, I had to challenge several notions and concepts that seemed to have no further advancement under their current paradigms or reference frameworks, which will hopefully find a burst of fresh air in my words to keep science moving forward.

However, we all need to recognize that there is so much more that we can learn about this experience that we call life and that our knowledge is a great responsibility that could be used for good or bad. Reasons why there will always be a race inside our consciousness between our hearts and our minds, which I hope for the well-being of our species that the hearts of future generations will win it.

Now, if you followed my steps through these pages, you probably already realized why the book is titled "**The Footprints of the Atoms.**" Because I honestly believe that this is the next frontier that science has to tackle to find answers to the origins of all living entities on planet Earth, regardless of how we choose to define it.

And there is a good reason: finding a path to its origins is virtually impossible without understanding how life works at the most basic level.

Now, chances are that once we start walking toward this new frontier, we will discover that there is everything that we cannot see beyond all the things we can see, that beyond all the small things, there is everything that is even smaller. And that beyond all the great things, everything is even greater. And that all of these things are united in only one of which every one of us is a living part of it, our universe.

Important Definitions

In all fields of science, it is essential to have clear definitions and consistency in the terms and concepts used to describe our reality's observed events and behaviors.

And it is not difficult to see that when we fail to reach a consensus within the context of our scientific observations, this is usually caused by some arbitrary assumptions or contradictory and ambiguous definitions at the core of our discussions. This level of disagreement, more often than not, indicates that we are trying to describe these events and behaviors from vague references. And if we fail to recognize this condition early in the game, we analyze the information at hand to reach conclusions that will probably be incorrect.

Therefore we can't underestimate the importance of having clear definitions and adherence to our language's grammatical rules when communicating scientific proposals. Otherwise, we run the risk of committing many logical and reasoning errors.

To this effect, I have collected some basic definitions of various terms and concepts that we need to have while reading this material. These are provided as a quick reference starting point. But the reader is encouraged to look for more definitions and resources as needed to facilitate the analysis and understanding of the materials reviewed.

Important Definitions

Abiogenesis[12]

The origin of life from nonliving matter.

A "theory" in the evolution of early life on Earth that states that organic molecules and subsequent simple life forms originated from inorganic substances.

According to the conventional hypothesis, the earliest living cells emerged due to chemical evolution on our planet billions of years ago.

Biogenesis[13]

The development of life from preexisting life. The synthesis of chemical compounds or structures in living organisms.

Causation[14]

Causation, or causality, is the capacity of one variable to influence another. The first variable may bring the second into existence or cause the second variable's incidence to fluctuate.

Empirical[15]

Originating in or based on observation or experience.

Relying on experience or observation alone, often without due regard for system and theory.

Capable of being verified or disproved by observation or experiment.

---

12 Abiogenesis Definition & Meaning - Merriam-Webster
13 Biogenesis Definition & Meaning - Merriam-Webster
14 What is causation? | Definition from TechTarget
15 Empirical Definition & Meaning - Merriam-Webster

Epistemic[16]

Of or relating to knowledge or knowing.

Evidence[17]

The available facts or information indicating whether a belief or proposition is true.

Evidence, broadly construed, is anything presented in support of an assertion. This support may be strong or weak.

The strongest type of evidence provides direct proof of the truth of an assertion. In contrast, the weakest type of evidence is merely consistent with a statement but does not rule out other contradictory assertions, as in circumstantial evidence.

Evidence is the foundation of all humankind's knowledge which makes possible the current state of technologies that has been developed up to this day in all fields of science. It is the irrefutable proof and the strongest criteria that we have to determine whether a given belief or proposition is true or false.

Existence[18]

The fact or state of living or having objective reality.

Existence is the ability of an entity to interact with physical or mental reality.

---

16 Epistemic Definition & Meaning - Merriam-Webster
17 Evidence Definition & Meaning | Dictionary.com
18 Existence - Wikipedia

IMPORTANT DEFINITIONS

The notion of existence refers to the act of actually being there and being able to interact through our physical senses with all that is real, as opposed to that which is merely imaginary.

Everything that exists is observable (including but not limited to our vision) and has direct or indirect evidence of its existence from the measurable effects they produce on other things.

Extraterrestrial[19]

In or coming from a place other than planet Earth.

Fact[20]

A truth known by experience or observation; something known to be true.

False[21]

Not true or correct; erroneous: a false statement.

Uttering or declaring what is untrue: a false witness.

Not faithful or loyal; treacherous: a false friend.

Tending to deceive or mislead; deceptive: a false impression.

Not genuine; counterfeit.

---

19 EXTRATERRESTRIAL | definition in the Cambridge English Dictionary
20 Fact Definition & Meaning | Dictionary.com
21 False Definition & Meaning | Dictionary.com

Hypothesis[22]

A hypothesis is a tentative assumption, an idea proposed for the sake of an argument so that it can be tested to see if it is true.

Information[23]

Information is the physical representation of thoughts created on any existing matter or medium of our observed reality, modifying its physical properties or characteristics through the use of energy.

Intelligence[24]

Capacity for learning, reasoning, understanding, and similar forms of mental activity; aptitude in grasping truths, relationships, facts, meanings, etc.

Manifestation of a high mental capacity. The faculty of understanding.

Knowledge[25]

Knowledge is the awareness and understanding we gain through reasoning and experience of a particular science, principle, art, or technique.

Life[26]

The conditions that distinguish animals and plants from inorganic matter include the capacity for growth, reproduction, functional activity, and continual change preceding death.

---

22 Hypothesis Definition & Meaning - Merriam-Webster
23 Footprints of the Atoms, P. 154
24 Intelligence Definition & Meaning | Dictionary.com
25 https://www.merriam-webster.com/dictionary/knowledge
26 life | Definition, Origin, Evolution, Diversity, & Facts | Britannica

Non-Science[27]

Something (such as a discipline) that is not a science. Of or relating to fields other than science.

A non-science is an area of study that is not scientific, especially one that is not a natural science or a social science that is an object of scientific inquiry. In this model, history, art, and religion are all examples of non-sciences.

Non-scientific[28]

Not involving or relating to science or scientific methods.

Organic[29]

Relating to, being, or dealt with by a branch of chemistry concerned with the carbon compounds of living beings and most other carbon compounds.

Origin[30]

The point at which something begins its course or existence. Origin applies to the things or persons from which something is ultimately derived and often to the causes operating before something comes into being.

Proposition[31]

Propositions deal with the connection between two existing/known concepts.

---

27 Nonscience Definition & Meaning - Merriam-Webster
28 Nonscientific Definition & Meaning | Britannica Dictionary
29 https://www.merriam-webster.com/dictionary/organic
30 https://www.merriam-webster.com/dictionary/origin
31 Difference Between Proposition & Hypothesis (sciencing.com)

Its primary purpose is to suggest a link between two concepts by formulating a possible answer to a specific scientific question where experiments cannot test and verify the connection between them.

As a result, propositions rely heavily on prior research, reasonable assumptions, and existing correlative evidence.

Propositions spur research in promising areas of scientific inquiry by asking a question that can support further speculation in hopes that evidence or experimental methods will be discovered to make it a testable hypothesis.

These are very useful in areas of study like sociology and economics, where valid hypotheses can rarely be made due to their complexity, and experimental tests would be prohibitively expensive or complicated.

They are also valuable in areas where little hard evidence remains, such as archaeological and paleontological studies in which only fragments of evidence have been discovered.

But because propositions do not rely on testable data; they only need to be convincing and internally consistent to appear valid and accepted, which makes them extremely difficult to disprove in the scientific context.

Pseudoscience[32]

Pseudoscience consists of statements, beliefs, or practices that claim to be scientific and factual but incompatible with the scientific method.

Pseudoscience is often characterized by contradictory, exaggerated, or unfalsifiable claims; reliance on confirmation bias rather than rigorous attempts at refutation; lack of openness to evaluation by other

---

32   https://en.wikipedia.org/wiki/Pseudoscience

experts; absence of systematic practices when developing hypotheses; and continued adherence long after the pseudo-scientific hypotheses have been experimentally discredited.

Science[33]

Science is the pursuit and application of knowledge and understanding of the natural and social world following a systematic, evidence-based methodology.

Scientific[34]

Based on or characterized by the methods and principles of knowledge and understanding of the natural and social world following a systematic methodology based on evidence.

Scientific Hypothesis[35]

A scientific hypothesis is a tentative answer to a scientific question consistent with the scientific analysis method made with limited evidence as a starting point for further investigation. It is a tentative assumption made to draw out and test its logical or empirical consequences.

Scientific Theory[36]

A scientific theory is an explanation of an aspect of the natural world that has been repeatedly tested and corroborated following the scientific method, using accepted protocols of observation, measurement, and evaluation of results. Where possible, theories are tested under

---

33 Our definition of science - The Science Council ~ : The Science Council ~
34 Scientific Definition & Meaning - Merriam-Webster
35 Hypothesis Definition & Meaning - Merriam-Webster
36 Scientific theory - Wikipedia

controlled conditions in an experiment. In circumstances not amenable to experimental testing, theories are evaluated through principles of abductive reasoning. Established scientific theories have withstood rigorous scrutiny and embody scientific knowledge.

True[37]

Being in accordance with the actual state or conditions; conforming to reality or fact; not false:

Something accurate, correct, and verifiable per facts, evidence, or reality.

Truth[38]

The body of real things, events, and facts.

The state of being the case.

The body of factual statements and propositions

The property (as of a statement) of being in accord with fact or reality

---

[37] True Definition & Meaning | Dictionary.com
[38] Truth Definition & Meaning - Merriam-Webster

Logic 101

This review is intended to provide a basic level of awareness of the most commonly used reasoning modes to draw conclusions, make predictions or construct explanations of our observed reality. It is provided so that readers have a quick reference and feel confident about the analysis and the conclusions presented in this work.

Now keep in mind that these basic guidelines and the examples presented here are not a comprehensive training on all aspects of logical reasoning. Since this will take a considerable amount of time and effort, and it is not the main scope of this work.

Nevertheless, readers are encouraged to research and learn any of the topics of logical reasoning necessary to complement, validate or refute the conclusions presented.

What is logic?

Logic (The Basics of Philosophy, 2008) is one of the many types of reasoning accepted in different science fields and almost every context of our day-to-day life. It is the science that we use to explain or represent a consistent argument about a particular topic.

Logic relies on an argument's basic form or structure to describe something that comes from clear reasoning. So calling something logical means it's based on good reasoning and sound ideas.

And as logical as it sounds, not many of us are familiar with logical reasoning concepts. It may be common sense to think that everyone should know this because we depend on it in our day-to-day endeavors for all decision-making purposes. Still, the fact is that many of us have never been exposed to its basic concepts, and we are unaware of them.

This lack of "logical awareness" more than often hinders our ability to differentiate a truth from a lie, and for most of our lives, we make use of our reasoning power without even noticing that we are committing lots of reasoning errors that deceive and lead us to believe many lies as if they were true.

And this is where knowing a little bit about logic can make a big difference since logical reasoning can explain and validate every truth through a series of rules of necessary relationships that help us differentiate between true and false.

The criterion used in logical reasoning to verify those relationships is the truth. So if the truth is true, logic and reason will necessarily describe the components that constitute that truth.

Nevertheless, logical reasoning can't identify what is true in the first place. And this is the case because logical reasoning can only proceed from truths already known to be true or from truths assumed to be true for logical testing purposes. Therefore logical reasoning explains how the things we know or presume to be true can be explained according to their necessary logical relationships, which helps us test and validate our knowledge and its corresponding assumptions.

So, suppose I present a critical analysis of several scientific concepts. In that case, it seems appropriate that I take the time to test its logical characteristics when needed since it will be a bad reasoning error to assume that what is being proposed is correct to find out otherwise later.

What is a logical argument?

If you use words to try to make a point to prove something or persuade someone of something, then you are making an argument (What is a logical Argument?, 2022), which is the process by which one explains how a conclusion was reached?

Generally, the opening statement used as the starting point to build any argument is usually a proposition.

Everyone argues their positions and may choose to do so in various manners. However, a logical argument is composed of one or multiple premises and a conclusion that follows certain guiding principles or procedures in hopes of arriving at the desired conclusion, whose ultimate goal is to present an idea that is both consistent and coherent.

The premises of an argument are the statements (one or many) containing the evidence presented (reasons) to determine the degree of truth of the conclusion. And these can be true or false and must have a valid logical relation to the argument's conclusion. These premises usually contain words like because, as, since, seeing that, given that, etc. While the conclusion is the statement that includes the point that the argument is trying to make or prove.

There is usually only one conclusion in a given argument. And it contains words like therefore, wherefore, so, thus, hence, accordingly, etc.

Conclusions can be true or false, depending on the argument's premises and how well these support its logical relationships.

Argument Examples

Let's look at a couple of simple argument examples so that we can recognize their premises and their conclusion:

1. If Socrates is a man, and all men are mortal, then Socrates is mortal.

    Premises:
    a. Socrates is a man.
    b. All men are mortal.

    Conclusion:
    c. Socrates is mortal.

This argument tries to prove that Socrates is mortal because evidence shows that Socrates is a man and that all men are mortal. Therefore if Socrates belongs to the group of men, and all of them are mortal, he has to be mortal too.

2. The streets are wet. Streets get wet when it rains. Therefore it must have rained.

    Premises:
    a. Streets are wet.
    b. Streets get wet when it rains.

    Conclusion:
    c. Therefore, it must have rained.

This argument tries to prove that it rained outside because evidence shows that the streets get wet every time it rains. Therefore, it must have rained outside.

3. Robert smokes. People who smoke get cancer. Therefore Robert will get cancer.

   Premises:
   a. Robert smokes.
   b. People that smoke get cancer.

   Conclusion:
   c. Therefore, Robert will get cancer.

This argument tries to prove that Robert will get cancer because he smokes, and a person who smokes gets cancer.

What is a logical analysis?

Logical analysis (The Basics of Logical Analysis, 2019) is a detailed examination of an argument's elements and structure, typically as a basis for discussion or interpretation used to recognize a good or a bad argument.

A logical analysis follows a consistent set of rules to evaluate arguments based on their premises and conclusions. And these rules are used to identify the type of reasoning mode used by any given argument.

Being able to differentiate between good and bad arguments improves critical thinking abilities and helps us make better decisions in our daily life.

The standard or reference used for evaluating any logical argument is the truth.

Valid arguments are considered to be true, while invalid arguments are deemed to be false.

Any argument's validity is determined based on its premises and logical relationship to its conclusion.

An invalid argument is considered bad reasoning, and one or more premises are usually false or do not relate to or support its conclusion.

It is crucial to remember that only premises that are part of an argument can be considered when evaluating the validity of a given argument. Premises that are not contained as part of an argument <u>should not be assumed to be part of the argument</u> and should not be considered during the analysis of the argument.

What is a fallacy?

A fallacy is a misconception resulting from a failure in logical reasoning, which renders an argument invalid, causing deception and mistaken beliefs. It is a trick or illusion in thoughts that often succeeds in obfuscating facts/truth (Logical Fallacies, 2022).

For example, if you find that a premise is not true or does not support the conclusion of an argument, then you have found a fallacy.

There are two main forms of fallacies, a formal fallacy and an informal fallacy.

Formal fallacies are defined as an error that can be seen within the argument's form. And informal fallacies refer to an argument whose proposed conclusion is not supported by the premises. Both of them create an unpersuasive or unsatisfying conclusion.

There are many kinds of fallacies in logical reasoning, but we will only review the ones relevant to our analysis.

Ambiguity/Equivocation[39]

The fallacy of equivocation uses misleading terms of more than one meaning without clarifying which definition is intended in the scenario.

An excellent example of this is the use of the word theory when used within the context of a statement meaning a hypothesis or not proven explanation.

Appeal to Consequences[40]

It occurs when the truthfulness of a statement or belief is decided by the consequences it would have. It's used, perhaps most commonly, in attempts to either support or refute a particular belief, such as the existence of a higher being.

The fact that a proposition leads to some unfavorable result does not mean it is false. Similarly, just because the proposition has good consequences does not suddenly make it true. In the case of good consequences, such an argument may appeal to an audience's hopes, which sometimes take the form of wishful thinking. In the case of bad consequences, the argument may instead play upon an audience's fears.

Argument from Fallacy[41]

The argument from fallacy, also known as the bad reasons fallacy, stems from the claim that because the reason(s) given for a certain conclusion are bad, the conclusion must also be incorrect. This

---

39 Ambiguity Fallacy (logicallyfallacious.com)
40 Appeal to Consequences - Definition and Examples - Fallacy In Logic
41 Argument from Fallacy (logicallyfallacious.com)

fallacy supposes that providing a lousy reason for a correct conclusion is impossible when in fact, it is possible to do so.

Example:

"Dogs are afraid of heights. Therefore dogs don't fly."

Though it may be true that dogs are afraid of heights, that is not why they do not fly.

Argument from Ignorance[42]

The assumption of a conclusion or fact based primarily on a lack of evidence to the contrary. Usually best described by, "absence of evidence is not evidence of absence."

Logical Forms:

X is true because you cannot prove that X is false.

X is false because you cannot prove that X is true.

Example:

To this day (at the time of this writing), science has been unable to create life from non-life; therefore, life must result from divine intervention.

The fact that we have not found a way to create life from non-life is not evidence that there is no way to create life from non-life, nor is it evidence that we will someday be able to; it is just evidence that we do not know how to do it. Confusing ignorance with impossibility (or possibility) is fallacious.

---

42 Argument from Ignorance (logicallyfallacious.com)

Blind Authority[43]

God of the gaps (or a divine fallacy) is a logical fallacy that occurs when believers invoke Goddidit (or a variant) to account for some natural phenomena that science cannot explain at the time of the argument.

Circular Reasoning[44]

Circular reasoning is a logical fallacy in which the reasoner begins with what they are trying to end with. It is an argument's defect where the premises are just as much in need of proof or evidence as the conclusion, and as a consequence, the argument fails to persuade.

In a circular reference fallacy, there is no reason to accept the premises unless one already believes the conclusion or the premises provide no independent ground or evidence for the conclusion.

> Logical Form:
> A is true because B is true;
> B is true because A is true.

In this type of argument, the premises do not meet the requirement of proving the conclusion; thus, it is a fallacy.

Example: The Bible is the Word of God because God tells us it is in the Bible.

Explanation: In this argument, the premise that "The bible is the word of god..." is as much in need of proof or evidence as the conclusion "God tells us it is in the bible" is. Therefore this argument

---

[43] Blind Authority Fallacy (logicallyfallacious.com)
[44] Circular Reasoning (logicallyfallacious.com)

fails to persuade because it does not provide any independent evidence for the conclusion.

Definist or Redefinition Fallacy[45]

The definist fallacy occurs when someone defines a concept in biased terms for the sake of argument. The person making the argument hopes their audience will accept the provided definition, making his position much easier to defend and more challenging to refute.

>Logical Form:
>
>A has definition X.
>X is harmful to my argument.
>Therefore, A has a definition of Y.

An excellent example of the definist fallacy could be a libertarian defining taxes as "the government stealing from the public." Though this term is loosely correct in that taxes do entail the government taking money from citizens, it places extreme negative bias on the idea by using the word "stealing" instead of "taking" or "withholding." This definition makes taxes seem inherently bad. Arguing that taxes are good and necessary would be very difficult if one assumed this word's meaning.

Therefore, it is essential not to accept definitions put forth by an arguer unless you researched the definition on your own and agree.

Disregarding Known Science[46]

---

45 Definist Fallacy (logicallyfallacious.com)
46 Fallacies | Internet Encyclopedia of Philosophy (utm.edu)

This fallacy is committed when a person makes a claim knowingly or unknowingly, disregarding well-known science that weighs against the claim. They should know better. This fallacy is a form of the Fallacy of Suppressed Evidence ignoring scientific facts that are inconvenient to a position.

This is better characterized as a cognitive bias or perhaps even a form of lying.

False Dilemma[47]

The false dichotomy fallacy occurs when the range of options being analyzed is oversimplified.

This line of reasoning fails by limiting the options of an argument to two when there are more options to choose from.

For example, "Life on Earth evolved from physical matter on Earth, or an intelligent designer created it."

In fairness, life on Earth could;

    A. has evolved in another part of the universe with adequate environmental conditions, then;

        1. Been brought to Earth (not created) by biological entities from these locations.

        2. Been seeded on Earth by asteroids from other parts of the universe.

    B. has been created by intelligent entities that genetically modified other species from other parts of the universe.

---
47 False Dilemma (logicallyfallacious.com)

C. Or it could have also been created on Earth with existing materials by an intelligent entity using means or methods unknown to us.

D. A combination of the above, etc., etc.

The false dilemma fallacy is often a manipulative tool designed to polarize the audience between two choices when more options could be available to prove a conclusion.

Incomplete Comparison[48]

An incomplete comparison occurs when two things are compared that are not related to make something more appealing than it is. This also happens when conclusions are made with incomplete information.

Non-Sequitur[49]

This type of fallacy is committed when the conclusion of an argument does not follow from its premises. It is when what is presented as evidence or reason is irrelevant or adds very little support to the conclusion of an argument.

This type of fallacy causes people to spread inaccurate information.

Logical Form:

Claim A is made.
Evidence is presented for claim A.
Therefore, claim C is true.

---

48 Incomplete Comparison (logicallyfallacious.com)
49 Non Sequitur (logicallyfallacious.com)

Example: Buddy Burger has the greatest food in town. Buddy Burger was voted #1 by the local paper. Therefore, Phil, the owner of Buddy Burger, should run for president of the United States.

Explanation: I bet Phil makes one heck of a burger, but it does not follow that he should be president.

Proving non-Existence[50]

This fallacy occurs when someone demands that one proves the non-existence of something instead of providing adequate evidence for the existence of that something.

Logically speaking, the proof of the existence of something must come from those who make the claims. And no evidence of the non-existence of something should come from those you are trying to convince in the first place.

> Logical Form:
>
> I cannot prove that X exists, so you prove it doesn't.
> If you can't, X exists.

Example #1:
God exists. Until you can prove otherwise, I will believe he does.

Explanation: There are decent reasons to believe in the existence of God, but "because the existence of God cannot be disproved" is not one of them.

---

[50] Proving Non-Existence (logicallyfallacious.com)

Questionable Cause[51]

Concluding that one thing causes another simply because they are regularly associated.

> Example:
>
> Every time I go to sleep, the sun goes down. Therefore, going to sleep causes the sun to set.

Of course, you probably sleep like everyone else when the sun goes down, but it is evident that this does not cause the sun to go down.

Reification[52]

Also called the Fallacy of Misplaced Concreteness, the reification fallacy occurs when an argument relies on an abstract concept as a concrete fact, when a hypothetical scenario or situation is referred to and treated as real.

In other words, it is the error of treating something that is not concrete, such as an idea, as tangible. A typical case of reification is the confusion of a model with reality: "the map is not the territory."

> Example:
>
> If you are open to it, love will find you.

Love is an abstraction, not a little fat-flying baby with a bow and arrow that searches for victims. Cute sayings such as this can serve

---

51 Questionable Cause (logicallyfallacious.com)
52 Reification (logicallyfallacious.com)

as bad advice for people who would otherwise try to find a romantic partner but choose not to, believing that this "love entity" is busy searching for their ideal mate.

Weak Analogy Fallacy[53]

The fallacy of weak analogy is a fallacy of weak induction. This means there is insufficient support from the premises to believe the conclusion rationally.

The weak analogy fallacy is committed when the premises of an argument presents an analogy or similarity between two things or situations. But the analogy is not strong enough to support the conclusion.

Example: "Guns are like hammers—they're tools with metal parts that could be used to kill someone.

And yet it would be ridiculous to restrict the purchase of hammers—so restrictions on purchasing guns are equally ridiculous." While guns and hammers share certain features, these features (having metal parts, being tools, and being potentially useful for violence) are not the ones at stake in deciding whether to restrict guns. Instead, we restrict guns because they can easily be used to kill large numbers of people at a distance. This is a feature hammers do not share—it would be hard to kill a crowd with a hammer. Thus, the analogy is weak, and so is the argument based on it.

---

53 Weak Analogy (logicallyfallacious.com)

Reasoning Modes

Reasoning (Spacey, 2018) is the process of using existing knowledge to draw conclusions, make predictions, or construct explanations. The three most commonly used reasoning modes (DeMichele, 2017) to present arguments about empirical observations are deductive, inductive, and abductive. And they can be differentiated by the order in which the Rule, the Cause, and the Effect of a given argument are located within its premises and the conclusion.

Note that in the following three reasoning modes examples, the Rule, the Cause, and the Effect are the same to enforce the necessary logical relationship between the premises and the conclusion of the arguments to avoid logical fallacies.

The RNA World Example

Deductive Reasoning
Deductive reasoning is used to identify an effect when we have a rule and a cause. The Rule and the Cause are part of the argument's premises, and the effect is the necessary conclusion. For example;

Premises
Rule - All abiogenesis hypotheses can't prove how life evolved at random from inert matter.
Cause - The RNA World Hypothesis is an abiogenesis hypothesis.

Conclusion
Effect - Therefore, the RNA world hypothesis can't prove how life evolves at random from inert matter.

The Rule and the Cause are part of the argument's premises in this example. And the Effect is its certain conclusion.

In this reasoning mode, the arguer claims that if the argument's premises are true (Rule and Cause), the conclusion (Effect) can't be false.

Suppose all abiogenesis hypotheses can't prove how life evolved from matter, and the RNA World Hypothesis is considered part of this hypothesis group. In that case, the necessary logical conclusion is that the RNA world hypothesis can't prove how life evolves from matter, either.

This effect is the necessary logical conclusion derived from this argument's premises.

A deductive reasoning argument can be disproved by disproving any one of its premises. For example, if one can prove that the RNA World Hypothesis is not an abiogenesis hypothesis, its conclusion will be false since it was derived from a false premise.

Inductive Reasoning

Inductive reasoning is used to form rules when we have a cause and its effect. The cause and the effect are part of the argument's premises, and the rule is a probable conclusion. It is probable, but it is not entirely certain.

Premises
Cause - The RNA World hypothesis is an abiogenesis hypothesis.
Effect - The RNA World hypothesis can't prove how life evolves from matter.

Conclusion
Rule - Therefore, all abiogenesis hypotheses can't prove how life evolved from matter.

In this example, the Cause and the Effect are part of the argument's premises, and the Rule is its probable conclusion but not a certain one.

In this type of reasoning mode, the arguer claims that if the argument's premises are true (Cause and Effect), it is improbable for the conclusion (Rule) to be false.

Although the RNA World Hypothesis is indeed a known abiogenesis hypothesis that can't prove how life evolved from matter, based on the information contained in these premises of the argument, it is not certain to conclude that all abiogenesis hypotheses can't prove how life evolved from matter since this information is not part of the premises.

Therefore, based on the information provided on the argument's premises, we can't conclude with certainty that all other abiogenesis hypotheses can't prove how life evolves from matter. So the rule is a probable conclusion, but not a certain one.

Note that although, up to this day, we know as a fact that all known abiogenesis hypotheses can't prove how life evolves from matter, this piece of information can't be considered part of the analysis of the argument since it is not part of its premises. And it is unknown, based on the premises, whether or not some of the existing abiogenesis hypotheses can prove how life evolved from inert matter.

An inductive reasoning argument can be disproved if an example that contradicts its conclusion can be found. For instance, if I were to find an abiogenesis hypothesis that can prove how life evolved from matter, then the conclusion would be disproved.

Abductive Reasoning

Abductive reasoning is used to identify a cause from a rule and an effect. The rule and the effect are part of the argument's premises, and the cause is a possible conclusion. It is possible, but it is not entirely certain.

Premises
Rule - All abiogenesis hypotheses can't prove how life evolved at random from inert matter.
Effect - The RNA world hypothesis can't prove how life evolves at random from inert matter.

Conclusion
Cause - Therefore, the RNA World Hypothesis is an abiogenesis hypothesis.

In this type of reasoning mode, the arguer claims the argument is unlikely to be false. If the premises are true, it is unlikely for the conclusion to be false.

In this type of argument, like in the inductive type, although, indeed, all abiogenesis hypotheses and the RNA World Hypothesis can't prove how life evolves from matter, it is not certain to conclude based on these premises that the RNA World Hypothesis is an abiogenesis hypothesis, because the argument does not provide any premises about whether or not the RNA World Hypothesis is or is not part of the group of the abiogenesis hypothesis. It may very well be part of another group of hypotheses.

Therefore based on the premises, we can't conclude with certainty that the RNA World hypothesis is an abiogenesis hypothesis. So its conclusion is a likely cause, but not a certain one.

Note that although we know as a fact that the RNA World hypothesis is an abiogenesis hypothesis, this can't be considered as part of the analysis of the argument since it is not part of its premises. Therefore based on the information provided in its premises, the abductive reasoning mode gives us a conclusion that is a likely cause but not a certain one.

An abductive reasoning argument is usually incomplete and does not have absolute information; it can be disproved by finding a flaw in the explanation or the evidence. For example, the above conclusion can be disproved if we can prove that the RNA Hypothesis does not belong to the group of abiogenesis hypotheses but a different group of hypotheses. Or if we can prove that there is one abiogenesis hypothesis that can demonstrate how life evolved from matter.

Of the three basic logical reasoning modes, abduction is less accurate because it does not give us certainty. But it can be helpful when you don't have adequate evidence from past events. And it is very useful when used to generate plausible ideas but not for proving them.

Abductive reasoning eliminates what is obviously not true and considers what is most likely to be true.

Abduction does not reason straight from the premises to the conclusion as Deduction and Induction do. Instead, it reasons by ruling out possible explanations until you are left with the most plausible one.

The Beans from the Bag

Here is a much simpler example that illustrates the three different reasoning modes:

Deductive Reasoning

> Premises
> Rule: All the beans from this bag are white.
> Cause: These beans are from this bag.
>
> Conclusion
> Effect: These beans are white.

If all beans from this bag are white, and I pull some of them, they must be white.

Notice that if we know as a fact that all beans in the bag are white, and we pull some beans from it, they will necessarily be white beans since there are no other types of beans in the bag.

Inductive Reasoning

> Premises
> Cause: These beans are from this bag.
> Effect: These beans are white.
>
> Conclusion
> Rule: All the beans from this bag are white.

If these beans are from this bag, and the beans are white, then all beans from the bag are white.

Notice that the beans I pulled from this bag are white does not necessarily prove that all the beans in the bag are white. There could be black or red beans in the bag too.

Abductive Reasoning

> Premises
> Rule: All the beans from this bag are white.
> Effect: These beans are white.
>
> Conclusion
> Cause: These beans are from this bag.

Notice that all the beans from this bag are white, and the ones I have at hand are white, too, only proves that it is likely that the beans I have at hand are from the bag. But they may be from a different bag too. Therefore the conclusion is likely, but not certain.

Bibliography

8 Tools and Techniques of Gene Manipulation. (2018, 12 23). Retrieved 10 12, 2022, from Explore Biotech: https://explorebiotech.com/tools-and-techniques-of-gene-manipulation/

abiogenesis. (n.d.). Retrieved 10 8, 2022, from Merriam-Webster: https://www.merriam-webster.com/dictionary/abiogenesis

About. (2022). Retrieved 10 9, 2022, from Discovery Institute: https://www.discovery.org/id/about/

Academy, K. (2022, 12 5). *Cellular Organelles and Structure*. Retrieved 2 5, 2023, from Khanacademy: https://www.khanacademy.org/test-prep/mcat/cells/eukaryotic-cells/a/organelles-article

All3DP, E. (2023, 2 3). *The 7 Main Types of 3D Printing Technology*. Retrieved 3 16, 2023, from All3DP: https://all3dp.com/1/types-of-3d-printers-3d-printing-technology/

Amit, S. (2021, 12 27). *What is Theory of Biogenesis?* Retrieved 10 12, 2022, from Fun Biology: https://www.funbiology.com/biogenesis-theory/

Anderson, R. (2015). *The Cosmic Compendium: The Big Bang and the Early Universe*. New York: The Rosen Publishing Group.

Andres, V. (2022). Richards J. Heuer Jr. (1927–2018): Searcher for Truth and the Means to Recognize It. *Studies in Intelligence, Vol 62, No. 3*, 4.

ASCII Code - The extended ASCII table. (2022). Retrieved 10 11, 2022, from ASCII Code.com: https://www.ascii-code.com/

Augustyn, A. (2022). *Nucleation*. Retrieved 11 16, 2022, from www.britanica.com: https://www.britannica.com/science/nucleation

Bailey, R. (2020, 1 24). *Cell Theory: A Core Principle of Biology*. Retrieved 10 12, 2022, from ThoughtCo: https://www.thoughtco.com/cell-theory-373300

Bioinformatics. (2022, 10 10). Retrieved 10 14, 2022, from Wikipedia: https://en.wikipedia.org/w/index.php?title=Bioinformatics&oldid=1115179526

Bortz, F. (2014). *The big bang theory; Edwin Hubble and the origins of the universe*. New York: The Rosen Publishing Group.

Britannica, T. E., & Rogers, K. (22, 12 1). *Eukatyote*. Retrieved 2 5, 2023, from Britannica: https://www.britannica.com/science/eukaryote

Brooker, R. J., Widmaier, E. P., & Graham, L. (2008, 12 2020). *Biology Seven Characteristics of Life | Science Facts*. New York: McGraw-Hill.

Byjus. (2022). *Doppler Effect - Definition, Formula, Examples, Uses, FAQs*. Retrieved 11 25, 2022, from BYJU'S: https://byjus.com/physics/doppler-effect/

Canetti, L., Dewes, M., & Shaposhnikov, M. (2012). Matter and antimatter in the universe. *New Jornal of Physics, 14* (1), 1-20.

Cells. (2022). Retrieved 10 11, 2022, from BYJUS: https://byjus.com/biology/cells/

Cherry, K. (2022, 3 1). *How Different Psychologists Have Evaluated Intelligence*. Retrieved 10 12, 2022, from Verywell Mind: https://www.verywellmind.com/theories-of-intelligence-2795035

Conley, C. (2022, 8 19). *6 Steps Scientific Method*. Retrieved 10 10, 2020, from Hartford Community College: https://harford.libguides.com/c.php?g=321391&p=2150319

Cooper, G. M. (2000). *The Origin and Evolution of Cells*. Sunderland (MA): Sinauer Associates.

Daley, J. (2016, 6 26). *Behold LUCA, the Last Universal Common Ancestor of Life on Earth*. Retrieved 10 12, 2022, from Smithsonian Magazine: https://www.smithsonianmag.com/smart-news/behold-luca-last-universal-common-ancestor-life-earth-180959915/

DeMichele, T. (2017, 6 15). *The Different Types of Reasoning Methods*. Retrieved 10 9, 2022, from FACT / MYTH: http://factmyth.com/the-different-types-of-reasoning-methods-explained-and-compared/

DeMichelle, T. (2021, 03 1). *You Can't Prove a Negative*. Retrieved 10 10, 2022, from Fact / Myth: http://factmyth.com/factoids/you-cant-prove-a-negative/

Deoxyribonucleic Acid (DNA) Fact Sheet. (2020, 8 24). Retrieved 10 11, 2022, from Genome.gov: https://www.genome.gov/about-genomics/fact-sheets/Deoxyribonucleic-Acid-Fact-Sheet

D'Onofrio, D. J., & An, G. (2010). A comparative approach for the investigation of biological information processing: An examination of the structure and function of computer hard drives and DNA. *Theoretical Biology and Medical Modelling*, 1-3.

Encyclopaedia Britannica, T. E. (2023, 1 6). *Nltosis*. Retrieved 3 10, 2023, from Brittanica: https://www.britannica.com/science/mitosis/additional-info#history

Explanatory power. (2022, 10 5). Retrieved 10 14, 2022, from Wikipedia: https://en.wikipedia.org/w/index.php?title=Explanatory_power&oldid=1114189546

Foster, J. (2031, 9 30). *What is a Biosphere in Ecology? Examples and Meaning in Biology*. Retrieved 10 25, 2022, from www.jotscroll.com: https://www.jotscroll.com/biosphere-examples-meaning-in-biology

Frequently Asked Questions. (n.d.). Retrieved 10 9, 2022, from Discovery Institute: https://www.discovery.org/id/faqs/

Graham, P., Koji, T., & Zach, W. (2022). Paraconsistent Logic. In P. Graham, *The Stanford Encyclopedia of Philosophy.* Metaphysics Research Lab, Stanford University.

Greshko, M. (2017, 1 18). *The origins of the universe, explained.* Retrieved 11 22, 2022, from www.nationalgeographics.com: https://www.nationalgeographic.com/science/article/origins-of-the-universe

Helmenstain, A. M. (2017, 3 11). *Crystal Chemicals.* Retrieved 11 16, 2022, from www.ThoughtCo.com: https://www.thoughtco.com/chemicals-for-growing-crystals-607651

Helmenstine, A. M. (2019, 7 27). *What are the elements of a good hypothesis?* Retrieved 10 10, 2022, from TougtCo: https://www.thoughtco.com/elements-of-a-good-hypothesis-609096

Helmenstine, A. M. (2020, 2 12). *Oxidation Definition and Example in Chemistry.* Retrieved 11 16, 2022, from www.Thoughco.com: https://www.thoughtco.com/definition-of-oxidation-in-chemistry-605456

Hirst, K. K. (2019, 6 3). *What Scientists Have Learned About Ancient Cave Paintings.* Retrieved 10 14, 2022, from ThoughtCo: https://www.thoughtco.com/cave-art-what-archaeologists-have-learned-170462

History.com Editors. (2019, 9 18). *Neanderthals.* Retrieved 10 14, 2022, from History: https://www.history.com/topics/pre-history/neanderthals

Hypotheses about the origins of life. (2016). Retrieved 10 9, 2022, from Khan Academy: https://www.khanacademy.org/science/ap-biology/natural-selection/origins-of-life-on-earth/a/hypotheses-about-the-origins-of-life

IOP. (2022). *The Big Bang.* Retrieved 11 22, 2022, from www.iop.com: https://www.iop.org/explore-physics/big-ideas-physics/big-bang

Jones, J. E. (2005, 12 20). *Kitzmiller vs. Dover Memorandum of Opinion[USCOURTS-pamd-4_04-cv-02688-12.pdf (govinfo.gov)].* Retrieved 10 9, 2022, from Govinfo.gov: https://www.govinfo.gov/content/pkg/USCOURTS-pamd-4_04-cv-02688/pdf/USCOURTS-pamd-4_04-cv-02688-12.pdf

Kumar, V. (2022, 08 21). *Origin Of The Universe: 7 Different Theories.* Retrieved 11 22, 2022, from https://www.rankred.com/origin-of-the-universe-different-theories/: https://www.rankred.com/origin-of-the-universe-different-theories/

Laham, K. (2021, 1 22). *Mitosis vs. Meiosis.* Retrieved 3 16, 2023, from Biology Dictionary: https://biologydictionary.net/mitosis-vs-meiosis/

Latham, K. (3032, 2 3+). *Prokaryotes vs. Eukaryotes.* Retrieved 23 2, 3033, from www.biologydictionary.net: https://biologydictionary.net/prokaryotes-vs-eukaryotes/

Law, S. (2011, 09 15). *You Can Prove a Negative.* Retrieved 10 10, 2022, from Psychology Today: https://www.psychologytoday.com/us/blog/believing-bull/201109/you-can-prove-negative

life - Evolution and the history of life on Earth | Britannica. (2022). Retrieved 10 28, 2022, from www.britanica.com: https://www.britannica.com/science/life/Evolution-and-the-history-of-life-on-Earth

Lim, A., & Dutfield, S. (2022, 2 1). *What is biology?* Retrieved 11 15, 2022, from www.livescience.com: https://www.livescience.com/44549-what-is-biology.html

Logical Fallacies. (2022). Retrieved 10 9, 2022, from Logical Falallcies: https://www.logicalfallacies.org/

Lotha, G. (2023, 1 6). *DNA Chemical Compound.* Retrieved 2 7, 2023, from Britannica: https://www.britannica.com/science/DNA

Margulis, L. (2022, 9 5). *life - Life on Earth | Britannica*. Retrieved 10 23, 2022, from www.britanica.com: https://www.britannica.com/science/life

Markouski, G. (2022, 12 16). *Information Theory*. Retrieved 1 24, 2023, from Britanica Encyclopedia: https://www.britannica.com/science/information-theory

mars.nasa.gov. (n.d.). *NASA Mars Exploration*. Retrieved 10 12, 2022, from mars.nasa.gov: https://mars.nasa.gov/

Mateuszi, K., & Sharma, S. (2020). CRISPR-Based Editing Techniques for Genetic Manipulation of Primary T Cells. *Methods and Protocols, 79*.

Matter, elements, and atoms | Chemistry of life (article) | Khan Academy. (n.d.). Retrieved 11 6, 2022, from www.khanacademy.org/: https://www.khanacademy.org/science/ap-biology/chemistry-of-life/elements-of-life/a/matter-elements-atoms-article

Mcleod, S. (2020, 5 1). *Karl Popper - Theory of Falsification*. Retrieved 10 10, 2022, from Simply Psychology: https://www.simplypsychology.org/Karl-Popper.html

Meyer, S. C. (2009, 5). *Signature in the cell*. Seatle, WA, USA: HarperCollins e-books.

Mischa. (2016, 5 25). *How a battery works*. Retrieved 11 19, 2022, from www.science.org.au: https://www.science.org.au/curious/technology-future/batteries

National Geographic, E. (2023, 2 8). *Conservation of Energy and Mass*. Retrieved 2 8, 2023, from National Geographic: https://education.nationalgeographic.org/resource/resource-library-conservation-energy-and-mass

Newton's Laws of Motion. (n.d.). Retrieved 10 10, 2022, from Vedantu: https://www.vedantu.com/physics/newtons-laws-of-motion

NIH, E. (2022, 7 7). *Is eye color determined by genetics?* Retrieved 2 7, 2023, from National Library of Medicine: https://medlineplus.gov/genetics/understanding/traits/eyecolor/

NOT SCIENCE. (2005, 12 21). *The Patriot News*, p. 1.

Nutrition, C. (2022, 8 3). *GMO Crops, Animal Food, and Beyond.* Retrieved 10 12, 2022, from FDA: https://www.fda.gov/food/agricultural-biotechnology/gmo-crops-animal-food-and-beyond

Our Definition of Science. (2022). Retrieved 10 9, 2022, from The Science Council: https://sciencecouncil.org/about-science/our-definition-of-science/

Panda, I. (2020, 6 11). *Big Bang Theory : Assumptions and Scientific Evidence Essay.* Retrieved 11 23, 2022, from www.ivypanda.com: https://ivypanda.com/essays/big-bang-theory-assumptions-amp-scientific-evidence/

Pat, B. (2021, 4 2). *What is an exoplanet?* Retrieved 10 12, 2022, from Exoplanets/NASA/Gov: https://exoplanets.nasa.gov/what-is-an-exoplanet/overview

Pauling, L. C. (2022, 10 18). *Periodic Table.* Retrieved 11 18, 2022, from www.britannica.com: https://www.britannica.com/science/periodic-table

Pray, L. A. (2008). *Molecular Events of DNA Replication | Learn Science at Scitable.* Retrieved 10 14, 2022, from Nature: http://www.nature.com/scitable/topicpage/major-molecular-events-of-dna-replication-413

Pyramids of Giza | National Geographics. (2017, 1 21). Retrieved 10 11, 2022, from History: https://www.nationalgeographic.com/history/article/giza-pyramids

Quantum Theory. (2022). Retrieved 10 10, 2022, from BYJUS: https://byjus.com/jee/quantum-theory/

Rafferty, J. P. (2020, 1 16). *Fossil Record*. Retrieved 2 16, 2023, from Britannica: https://www.britannica.com/science/fossil-record/additional-info#history

Rafferty, J. P. (2023, 2 12). *Just How Old Is Homo sapiens?* Retrieved 2 12, 2023, from Britannica: https://www.britannica.com/story/just-how-old-is-homo-sapiens

Rhee, G. (2013). *Cosmic dawn: The search for the first stars and galaxies.* New York: Springer Science.

Rocke, A. J. (2022). *Chemistry*. Retrieved 11 18, 2022, from www.britannica.com: https://www.britannica.com/science/chemistry

Roger, K. (2022, 12 1). *Eukaryote*. Retrieved 2 5, 2023, from Brittanica: https://www.britannica.com/science/eukaryote

Rogers, K. (2023, 2 15). *Abiogenesis - Biology*. Retrieved 3 5, 2023, from Britannica: https://www.britannica.com/science/abiogenesis

Schumaker, A. (2021, 11 5). *Biological Information Processing*. Retrieved 2 5, 2023, from Karlsruhe Institute of Technology: https://www.nacip.kit.edu/327.php#:~:text=Biology,with%20other%20cells%20and%20organisms.

Schupback, J. N. (2016). Competing Explanations and Explaining Away Arguments. *Theology and Science*, 256-267.

Science, G. (2017, 8 17). *Branches of Life Sciences: 75+ branches and their meanings?* Retrieved 10 26, 2022, from www.golifescience.com: https://www.golifescience.com/life-sciences-branches/

Scoville, H. (2019, 7 1). *The Evolution of Eukaryotic Cells*. Retrieved 2 9, 2023, from ThoughtCo.: https://www.thoughtco.com/the-evolution-of-eukaryotic-cells-1224557

Segre, J. (2022, 5 24). *Virus*. Retrieved 11 12, 2022, from www.genome.gov: https://www.genome.gov/genetics-glossary/Virus

Shikha, S. (2022, 7 19). *Origin of Life: Theories, Evidence, Proof, and Examples*. Retrieved 23 2022, 10, from www.embibe.com: https://www.embibe.com/exams/origin-of-life/

Siegel, E. (2016, 6 30). *Where id the Big Bang Happened?* Retrieved 11 22, 2022, from www.forbes.com: https://www.forbes.com/sites/startswithabang/2016/07/30/ask-ethan-where-did-the-big-bang-happen/

Siegmund, D. O. (2023, 1 27). *Probability Theory, Mathematics*. Retrieved 2 16, 2023, from Britannica: https://www.britannica.com/science/probability-theory

Simran. (2021, 12 29). *What is Thought? Psychology of Thoughts*. Retrieved 4 2023, 1, from mantracare.com: https://mantracare.org/therapy/issues/what-is-thought/

Slamecka, V. (2023, 16 1). *Britannica*. Retrieved 2 5, 2023, from Information Processing: https://www.britannica.com/technology/information-processing

Smith, M. (2022, 10 11). *Genetic Engineering*. Retrieved 10 12, 2022, from Genome.gov: https://www.genome.gov/genetics-glossary/Genetic-Engineering

Somma, M. (2021, 10 20). *What Are the Flst Eukayotic Fossils?* Retrieved 2 16, 2023, from Sciencing: https://sciencing.com/first-eukaryotic-fossils-8163415.html

Spacey, J. (2018, 4 28). *7 Types of Reasoning*. Retrieved 10 9, 2022, from Simplicable: https://simplicable.com/new/reasoning

Stanford, C., Allen, J. S., & Anton, S. C. (2009, 10 27). *Biological anthropology : the natural history of humankind*. Upper Saddle River : Pearson/Prentice Hall, 2009. 2nd ed.

The 1st Amendment of the U.S. Constitution. (2022). Retrieved 10 9, 2022, from National Constitution Center: https://constitution-center.org/the-constitution/amendments/amendment-i

The Basics of Logical Analysis. (2019, 9 25). Retrieved 10 9, 2022, from LibreTexts: https://human.libretexts.org/Bookshelves/Philosophy/Fundamental_Methods_of_Logic_(Knachel)/1%3A_The_Basics_of_Logical_Analysis

The Basics of Philosophy. (2008). Retrieved 10 9, 2022, from Philosophy Basics: https://www.philosophybasics.com/branch_logic.html

The Human Genome Project. (2014, 4 4). Retrieved 3 10, 2023, from NIH: https://www.genome.gov/about-nhgri/Policies-Guidance/Copyright

The Next Stage of Evolution. (n.d.). Retrieved 10 11, 2022, from Futurism: https://futurism.com/the-next-stage-of-evolution-how-will-the-human-species-evolve

The Scientific Method. (2022). Retrieved 10 9, 2022, from Khan Academy: https://www.khanacademy.org/science/biology/intro-to-biology/science-of-biology/a/the-science-of-biology

Top Differences Between Darwinism And Neo-Darwinism. (2019). Retrieved 10 10, 2022, from BYJUS: https://byjus.com/neet/difference-between-darwinism-and-neo-darwinism/

Understanding Science. (2022, 4 19). Retrieved 10 14, 2022, from UC Museum of Palenotology: https://undsci.berkeley.edu/the-philosophy-of-science/

Uuganbold, T. (2007). Business and competitive analysis: effective application of new and classic methods. *Academia*, 1-22.

Valchanov, I. (2021, 10 20). *Measuring Explanatory Power with the R-squared*. Retrieved 3 7, 2023, from 365 Data Science: https://365datascience.com/tutorials/statistics-tutorials/r-squared/

BIBLIOGRAPHY

Vedantu. (2022). *Cell Structure and Function*. Retrieved 12 8, 2022, from www.vedantu.com: https://www.vedantu.com/biology/cell-structure-and-function

Ware, L. (2020, 5 29). *Teapots and unicorns: absence of evidence is not evidence of absence*. Retrieved 10 10, 2022, from Evidently Cochrane: https://www.evidentlycochrane.net/teapots-and-unicorns-absence-of-evidence-is-not-evidence-of-absence/

Was God an Alien? | Unveiled. (2022). Retrieved 10 14, 2022, from MSN: https://www.msn.com/en-us/video/animals/was-god-an-alien-unveiled/vi-AA12JH5I?category=foryou

What is a logical Argument? (2022). Retrieved 10 9, 2022, from Study.com: https://study.com/academy/lesson/logical-argument-definition-parts-examples.html

What is Russell's teapot? (2022, 1 4). Retrieved 10 10, 2022, from GotQuestions: https://www.gotquestions.org/Russells-teapot.html

What is the human body made of? (2020, 08 28). Retrieved 10 11, 2022, from BBC Science Focus Magazine: https://www.sciencefocus.com/the-human-body/what-is-the-human-body-made-of/

Zimmerman, A. J. (2019, 2 10). Retrieved 11 18, 2022, from https://www.thoughtco.com/what-is-physics-2699069

Zimmerman, A. J. (2020, 1 8). *Einstein's Theory of Relativity*. Retrieved 11 26, 2022, from ThoughtCo.: https://www.thoughtco.com/einsteins-theory-of-relativity-2699378

About the Author

Alfredo Archilla is a UPR-RUM graduate with a BSEE. From the early stages of his professional career, he has been fascinated by the ability that we have to transfer the decision-making process that "runs" within the boundaries of our brains into today's hardware and software intelligent devices capable of emulating our thought processes.

After being exposed for almost 40 years to all sorts of technologies to deliver automated solutions to the consumer electronics, automotive, aviation, and medical device industries, he developed an interest in understanding our logical decision-making process, which is the foundation used to write this book.

He is retired and lives in Charlotte, NC, with his lovely wife Carmen, where he spends most of his time thinking and writing about our observed reality.

Dedication

To the loving memory of our dear friend Nemo.

Nov. 24, 2018

www.ingramcontent.com/pod-product-compliance
Lightning Source LLC
Chambersburg PA
CBHW041920240526
45473CB00038B/2878